Biosynthesis
of
Acetate-derived Compounds

Biosynthesis
of
Acetate-derived Compounds

N. M. PACKTER

Department of Biochemistry
University of Leeds

A Wiley-Interscience Publication

JOHN WILEY & SONS
London · New York · Sydney · Toronto

Library of Congress Catalog card No. 72-8609

ISBN 0 471 65745 X

Printed in Great Britain by
J. W. Arrowsmith Ltd.,
Winterstoke Road, Bristol BS3 2NT

To Jacqueline

Preface

Progress in almost all aspects of lipid structure, metabolism and function has expanded rapidly in the past decade. Indeed interest in these topics overlaps a great deal and helps to swell the large number of research workers who study these interrelated fields. The knowledge that is gained is vital for a true understanding of the functioning of the living cell and of the sub-cellular organelles of which it is comprised.

This book is intended for undergraduates taking an Honours course in Biochemistry or related disciplines and for postgraduates who are interested in the general field of Lipid Biochemistry. It is assumed that these students possess a basic knowledge of the fundamental principles of Biochemistry. The depth of coverage is designed to give them a sound grasp of the topics dealt with. It is aimed at filling the fairly large gap that exists between the limited material available in standard texts and the very detailed expositions found in specialized reviews. Moreover, it is hoped that the description of the chemistry of the natural products involved will help the Biochemist while the emphasis on enzymological aspects should also assist the student who does not have a deep biochemical training.

The text deals essentially with areas of metabolism concerned with the biosynthesis of fatty acids in various tissues and organisms, lipids, phenols, sterols and other terpenoids of major significance, all of which derive from acetyl coenzyme A. A prime consideration throughout has been an attempt to interrelate the many anabolic pathways to which acetyl coenzyme A is subjected. Each chapter contains introductory comments on the structure and function of these compounds and some measure of historical review which permits the results of recent research to be placed in perspective. The material presented is based in the first instance on the content of lectures I give to undergraduates at Leeds. Reference is often made by name to the Head of a Laboratory or the principal investigator(s) concerned but most topics are simply referred to individually by a numbered sequence in the text. The relevant references (up to mid-1972) are given at the end of each chapter; detailed descriptions of experimental work have not been presented and should be sought, if required, in the original papers.

I would like to thank my colleagues who helped me on various occasions and especially Dr. Sheltawy who read and commented upon Chapter 5. However, my greatest debt is to my wife who showed much patience and forbearance during the period of writing (rather prolonged) and who typed the rough draft and manuscript with extreme efficiency and skill.

Acknowledgements

I wish to thank the following Authors and Publishers for their kind permission to reproduce data from tables which appear in the text as Tables 3.1 and 3.2 [Dr. S. J. Wakil and the American Society of Biological Chemists Inc. (*Journal of Biological Chemistry*)]; Table 4.1 [Dr. J. M. Lowenstein and the Biochemical Society (*Biochemical Journal and Biochemical Society Symposium*, No. 27)]; Table 4.2 [Dr. N. B. Myant and the Biochemical Society (*Biochemical Journal*)]; Table 5.1 [Dr. G. V. Marinetti and Elsevier Publishing Company (*Comprehensive Biochemistry*, Vol. 18)] and Table 8.1 [Professor T. W. Goodwin and the Biochemical Society (*Biochemical Journal*)].

Leeds   N. M. PACKTER
August, 1972

Contents

CHAPTER 1

Introductory Chapter

The Role of Acetyl-CoA in Metabolism

Many metabolites occupy central positions in intermediary metabolism, e.g. α-oxo acids that link the tricarboxylic acid cycle with amino acid metabolism, and glucose 6-phosphate that enters many diverse pathways. However, if one substance may be considered to have pride of place within the hierarchy of an integrated metabolism in living cells, this would un-doubtedly be acetyl-coenzyme A (acetyl-CoA, (I)). This special role is due

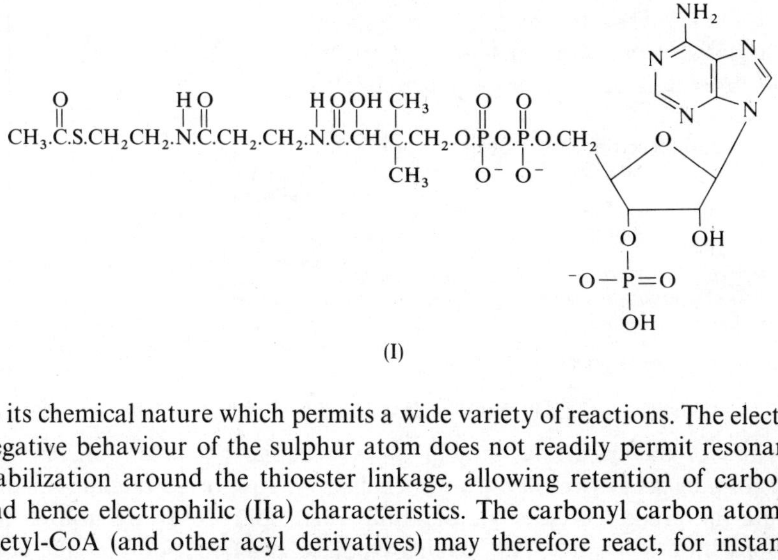

(I)

to its chemical nature which permits a wide variety of reactions. The electro-negative behaviour of the sulphur atom does not readily permit resonance stabilization around the thioester linkage, allowing retention of carbonyl and hence electrophilic (IIa) characteristics. The carbonyl carbon atom of acetyl-CoA (and other acyl derivatives) may therefore react, for instance, with thiol groups, water or amino substituted substrates in acyl transfer reactions in which coenzyme A is released. In addition, the α-carbon atom displays nucleophilic properties (IIb) as shown in the condensation of acetyl-CoA with oxaloacetate or acetoacetyl-CoA to form citrate or hydroxy-

(IIa)   (IIb)

1

methylglutaryl-CoA, and carboxylation with CO_2-biotinylenzyme (Chapter 2) to give malonyl-CoA.

Virtually all foodstuffs yield acetyl-CoA within the mitochondria as their end-product during catabolism. Its principal fate in aerobic organisms lies in reaction with oxaloacetate as acceptor molecule and entry into the tricarboxylic acid cycle where it is ultimately oxidized to CO_2 and H_2O with the generation of 'biochemical energy' as ATP. Metabolites formed in this cycle also serve as substrates for the synthesis of many building blocks for the macromolecules needed by the cell (Scheme 1.1). Operation of the glyoxylate ($HO_2C.CHO$) cycle,[1] that is reactions catalysed by isocitrate lyase (EC 4.1.3.1) and malate synthase (EC 4.1.3.2) together with some of the

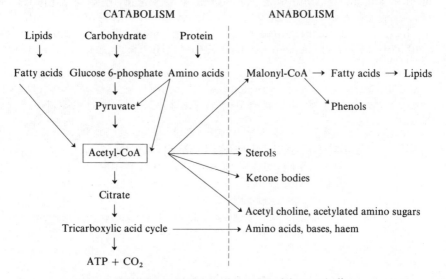

Scheme 1.1   The central role of acetyl-CoA in metabolism

enzymes of the tricarboxylic acid cycle, occupies a similar function for microorganisms grown on C_2 units (such as ethanol or acetate) and plants during the process of seed germination when lipid stores are utilized for carbohydrate synthesis. The malate synthase reaction catalyses the condensation of acetyl-CoA with the carbonyl (aldehyde) group of glyoxylate to give malate plus coenzyme A in a sequence that is analogous to citrate formation. This cycle acts as a means of restoring C_4 acids or indeed forming them *de novo* and also effectively oxidizes two molecules of acetyl-CoA to succinate *via* the intermediacy of isocitrate and malate (equations (1)–(3)):

$$\text{Isocitrate} \rightarrow \text{succinate} + \text{glyoxylate} \tag{1}$$

$$\text{Acetyl-CoA} + \text{glyoxylate} + H_2O \rightarrow \text{L-malate} + \text{CoA} \tag{2}$$

$$\text{Isocitrate} + \text{acetyl-CoA} + H_2O \rightarrow \text{succinate} + \text{L-malate} + \text{CoA} \tag{3}$$

The succinate thus formed may then be converted into oxaloacetate and eventually into glucose or amino acids. The key enzymes of this cycle, isocitrate lyase and malate synthase, are absent from animal tissues which are therefore incapable of performing a net synthesis of glucose from acetyl-CoA (or fatty acids).

With regard to anabolic sequences, the endoplasmic reticulum and cytosol (that is the non-membranous portion of the cell) contain enzymes of great synthetic capability that utilize the reactive C_2 unit of acetyl-CoA directly together with NADPH and an energy source for the synthesis of vital lipid components required by the membranous structures of the cell. In addition, fungi and plants in particular possess specific enzymes responsible for the formation of phenols and isoprenoid-derived products in a manner that resembles fatty acid and terpenoid synthesis. Acetyl-CoA possesses even greater versatility in its action as acetyl donor during the production of acetyl choline, an important nerve impulse transmitter, and acetylated amino sugars and other metabolites. N-Acetylglucosamine, for instance, is an essential constituent of fungal and bacterial cell-walls and of many vital glycoproteins including fibrinogen, blood group substances (secreted proteins) and thyroglobulin.

Furthermore, acetyl-CoA fulfils a very important non-substrate role as an allosteric effector (see below) for many regulatory enzymes concerned with this central region of metabolism. These enzymes exert great influence on the total flow and distribution of C_3 and C_4 acids and acetyl-CoA through catabolic and anabolic pathways.

The operation of the tricarboxylic acid cycle inevitably removes precursors for the biosynthetic functions of the cell. Transfer of acetyl-CoA and oxaloacetate from the mitochondria as citrate and eventual generation of the latter into pyruvate also results in the loss of the catalytic C_4 dicarboxylic acids from this cycle. It is essential that the rate of flow is maintained by means of an anaplerotic sequence[2] that supplies additional material. In animal tissues[3] and yeast[4] this role is filled by pyruvate carboxylase (EC 6.4.1.1), a biotin-containing enzyme, that is dependent upon acetyl-CoA for activity and replenishes the supply of oxaloacetate. A similar function is displayed by acetyl-CoA on activation of phosphoenolpyruvate carboxylase (EC 4.1.1.31) in some bacteria;[5] again a C_3 acid is converted into oxaloacetate under conditions in which acetyl-CoA is readily available. Activation of pyruvate carboxylase is also critical for the process of gluconeogenesis which arises *via* the intermediacy of oxaloacetate and phosphoenolpyruvate. This requirement for pyruvate is favoured by the inhibitory action of acetyl-CoA on pyruvate dehydrogenase (EC 1.2.4.1).

The material covered in this text attempts to deal with some of the biosynthetic features mentioned but discussion will be limited to aspects related to fatty acids, lipids, acetate-derived phenols, sterols and other terpenoids.

Energy Requirements

Synthetic processes in general utilize the same substances as precursors as the products liberated during catabolic reactions (or they may be provided in the diet) but there is an additional requirement for energy in order to overcome the unfavourable thermodynamic equilibria involved. Biosynthesis of metabolites is rarely accomplished by the direct reversal of the degradative pathway. In the latter case, larger molecules often supplied in the diet are mobilized into smaller, biochemically simpler products with the release of free energy, which is converted into chemical energy in the form of ATP. In connection with the formation of fatty acids, their derived lipids and certain phenolic products, condensation between the C_2 units concerned is only achieved after initial usage of ATP and bicarbonate during the carboxylation of acetyl-CoA to yield malonyl-CoA. The energy for the synthetic process is ultimately derived from the hydrolysis of ATP in a reaction catalysed by acetyl-CoA carboxylase (EC 6.4.1.2). The CO_2 released during the condensation between acetyl (or acyl) and malonyl thioester residues encourages C—C bond formation, thereby favouring reaction in the direction of synthesis.[6]

Regulation of Metabolic Pathways

Important considerations in any discussion of metabolic regulation are the consequences arising from the structural organization of membranes and cofactors within the cell. NADPH, the specific cofactor required for fatty acid (and sterol) synthesis, is formed entirely in the course of reactions occurring in the cytosol, with a resulting lack of competition for nucleotides between this sequence and the mitochondrial degradative pathway that utilizes NAD. A more elaborate modification of this compartmentation is found with the functional 4'-phosphopantetheine moiety that is used as acyl carrier during fatty acid synthesis. Acyl substrates are attached by thioester linkage to this cofactor which is itself bound to a small protein rather than an adenine nucleotide as in coenzyme A.

A more sophisticated but equally general mode of regulation of cellular economy is exerted by the presence of regulatory enzymes. These are particularly susceptible to control and occur at the beginning or end or branchpoint of metabolic sequences. They have rates that are considerably lower than the remaining enzymes in the pathway concerned and essentially act irreversibly; they therefore govern the overall rate. These enzymes uniquely possess properties that permit profound reversible modulation of their activity by stimulators and inhibitors by means of a phenomenon known as allosteric regulation.[7,8] Small molecule effectors bind at specific regulatory (non-catalytic) sites causing conformational changes in the quaternary structure, that is the overall three-dimensional shape, often by aggregation of the enzyme protomer or the reverse. This results in an alteration of the

K_m value (concentration of substrate at half-maximum saturation) thereby affecting the affinity for the substrate-binding site or of the v_{max} (maximum rate) of the reaction. The activity of key enzymes in metabolism is greatly modified by these means with a resultant coordination in cellular function. A characteristic sigmoidal curve is obtained instead of the more typical hyperbolic curve when enzyme activity is plotted against concentration of ligand (Figure 1(a)). This is due to a cooperative effect when conformational changes are slow compared with other stages in the reaction. Similarly, stimulation of the rate of a reaction at low substrate concentration may occur when the bonds within a polymeric unit are weakened in the presence of a metabolite which affects a regulatory sub-unit, thereby reducing the cooperative effect. Inhibition may be caused in a complementary manner by the activity of a different effector. The v_{max} remains unchanged and the enzyme is therefore released from inhibition at high substrate concentration (Figure 1(b)). An example of this type of system is found with aspartate transcarbamylase (aspartate carbamoyltransferase, EC 2.1.3.2), the initial enzyme

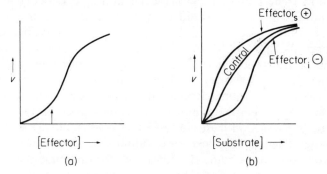

Figure 1.1   Representation of allosteric modulation of enzymic activity.
(a) Stimulation of rate of reaction by aggregation of monomers into the polymeric active enzyme. The arrow marks the concentration of effector above which the rate is rapidly increased in the presence of adequate substrate.
(b) Stimulation of rate over the control value at low substrate concentration by weakening bonds between the monomeric units within the polymeric form of the enzyme in the presence of effector$_s$.
Inhibition occurs in the presence of effector$_i$

engaged in pyrimidine biosynthesis and ultimately in the formation of the nucleotide triphosphates, UTP and CTP. This enzyme is strongly inhibited by cytidine nucleotides, especially CTP, and stimulated by ATP[9] in a process which eventually ensures a correct balance of purine and pyrimidine nucleotides.

Thus it can be seen that an immediate and fine control over the stringent metabolic demands dictated by the cell may prevail either by enhancement or feed-back inhibition (often by an end-product) of key enzymes in response to a threshold level of the effector in question. The action of many hormones, among others insulin and adrenaline, is partially mediated by their allosteric effect on vital regulatory enzymes.

Longer term effects that are brought about by changes in cell conditions are controlled by the concentration of enzyme protein. This, in turn, is dictated by similar allosteric regulation on enzyme synthesis at the genetic level, in the formation of the appropriate mRNA, as a result of induction or repression.[10] A cytoplasmic repressor protein coded for by the regulator gene (at least in bacterial systems) is bound selectively to the operator region in a strand of DNA and prevents transcription of the genetic unit concerned, and hence expression of the enzyme structural gene(s). Its ability to bind to this area is modified on attachment of a negative effector when induction of enzyme synthesis is gained, as the repressor is inactivated and binding is prevented, or by a positive effector (corepressor) that enhances blockage of the operator region. The effector molecule is often related structurally (or may indeed be identical) with the initial substrate or end-product of metabolic pathways and therefore acts as a physiological signal in the control of enzyme synthesis.

Bacteria have proved particularly useful as the source of experimental material for this type of study. In these organisms, enzymes belonging to many pathways are coordinately induced or repressed since their individual genes exist as gene clusters within the chromosome.[11] Their functional relationship is retained in these clusters as an operon where they are regulated together. There is some evidence to support the identity of gene clusters in eukaryotic organisms[12] and the concept of the operon may possibly be valid. Indeed, functional relationships have often been observed between enzymes in a common sequence including tightly integrated systems in the mitochondria and also in the cytosol. This might indicate the existence of constant proportion enzyme groupings that retain similar activity, relative to each other, despite large changes in their actual activity in different species or tissues.[13]

A more advanced evolutionary stage in this respect is exemplified *par excellence* in fatty acid synthetase from animal tissues and yeasts. These enzymes contain at least seven active proteins held together by non-covalent bonds. The organization of the genetic units responsible for the synthesis of the individual proteins of the yeast enzyme has recently been examined in mutant studies by complementation analysis.[12] Conservation of functionally related genes occurs on the yeast genome, indicating that they are coordinately transcribed prior to translation and assembly of the resultant proteins into the multienzyme complex. The properties of yeast and animal synthetases will now be described in detail in the following chapter.

REFERENCES

1. Kornberg, H. L. and Krebs, H. A., *Nature (London)*, **179,** 988 (1957)
2. Kornberg, H. L. In *Essays in Biochemistry*, Vol. 2, p. 1. Ed. by Campbell, P. N., and Greville, G. D., Academic Press Inc., London and New York, 1966
3. Utter, M. F., and Keech, D. B., *J. Biol. Chem.*, **238,** 2603 (1963); Keech, D. B., and Utter, M. F., *J. Biol. Chem.*, **238,** 2609 (1963)
4. Gailiusis, J., Rinne, R. W., and Benedict, C. R., *Biochim. Biophys. Acta*, **92,** 595 (1964)
5. Canovas, J. L., and Kornberg, H. L., *Biochim. Biophys. Acta*, **96,** 169 (1965)
6. Lynen, F., *Biochem. J.*, **102,** 381 (1967)
7. Monod, J., Changeux, J. P., and Jacob, F., *J. Mol. Biol.*, **6,** 306 (1963)
8. Monod, J., Wyman, J., and Changeux, J. P., *J. Mol. Biol.*, **12,** 88 (1965)
9. Gerhart, J. C., and Pardee, A. B., *J. Biol. Chem.*, **237,** 891 (1962)
10. Jacob, F., and Monod, J., *J. Mol. Biol.*, **3,** 318 (1961)
11. Ames, B. N., and Martin, R. G., *Annu. Rev. Biochem.* **33,** 235 (1964)
12. Kühn, L., Castorph, H., and Schweizer, E., *Eur. J. Biochem.* **24,** 492 (1972)
13. Greville, G. D. In *Citric Acid Cycle*, p. 52. Ed. by Lowenstein, J. M., Marcel Dekker, New York and London, 1969

CHAPTER 2

Biosynthesis of Fatty Acids in Animals and Yeasts

INTRODUCTION

Fatty acids are universal components of living matter and are readily synthesized by animals, plants and microorganisms. The common fatty acids in Nature are families of straight-chain compounds containing an even number of carbon atoms terminating in a carboxyl group; they may be saturated or unsaturated but usually do not contain more than four double bonds. The most widespread are palmitic ($C_{16:0}$), stearic ($C_{18:0}$) and oleic ($C_{18:1}$) acids. (The number preceding the colon designates the number of carbon atoms in the chain, whereas that following the colon refers to the number of double bonds present.) Shorter-chain acids ranging from C_4 to C_{12} are synthesized in the mammary gland and are present in milk; fatty acids with longer chain-lengths are present in animal tissues, especially the brain and nervous system, and in certain bacteria. More unusual derivatives possess acetylenic bonds, branched methyl groups and the related cyclopropane rings but these tend to occur in plants and microorganisms. Fatty acids containing odd numbers of carbon atoms are found universally.

Fatty acids seldom occur free in Nature but are generally present as components of more complex lipids in which they are bound covalently to various alcohols, e.g. in ester linkage to glycerol giving rise to triglycerides and phospholipids or in an amide bond with sphingosine (sphing-4-enine) which results in sphingomyelins and cerebrosides (Chapter 5). These products play important and diversified roles within the architecture of the cell and in its metabolism. Lipids and fatty acids form a heterogeneous collection of metabolically vital natural products. They are characteristically insoluble in water because of their hydrophobic nature, but readily soluble in many organic solvents. The constituent fatty acids may be isolated after saponification whereby they are converted into their water-soluble salts (soaps), followed by acidification and extraction into organic solvents.

Unsaturated acids are invariable components of lipids. They have melting points that are considerably lower than the corresponding saturated derivative and indeed this feature is essential for their physiological function in lipoprotein membranes surrounding the cell or subcellular organelles. The low melting point confers many desirable aspects of fluidity, and the degree of unsaturation among the constituent fatty acids of membrane phospholipids, for instance, controls their surface properties including that of permeability. The *cis* configuration present in these acids is extremely prevalent and arises through the specificity of the desaturase enzymes concerned in their formation.

Acids of varying degrees and geometry of unsaturation may be readily separated from each other and from saturated analogues by chromatographic treatment of their methyl esters on thin-layer plates of kieselgel or columns of silicic acid impregnated with silver nitrate (argentation chromatography).[1] Methyl esters of saturated acids (or lipids containing these groups) migrate at a normal rate in the usual solvent systems consisting of light petroleum–ether mixtures whereas the unsaturated derivatives are retarded; the greater the degree of unsaturation the lower the R_F value obtained. The reduced mobility is related to the complexing of the double-bond region with silver ions. Esters of hydroxy acids may also be separated by this means. The individual classes may be further resolved into their component fatty acids by application of gas–liquid chromatography.

Early Developments on the Biosynthesis of Fatty Acids

Studies on the formation of these metabolites were initiated during the classical work of Rittenberg and Bloch[2] who showed that both isotopes of doubly-labelled sodium acetate (CD_3 $^{13}COONa$) were incorporated into fatty acids and cholesterol in rats and mice. Degradation of the fatty acids indicated that the labelled atoms were distributed at alternate positions along the chain. Later work confirmed that the radioactive sodium [^{14}C] acetate was incorporated into fatty acids by liver slices and extracts and then also in mammary gland, heart, lung and other tissues. The principal fatty acids produced possessed even numbers of carbon atoms and were the C_{16} and C_{18} homologues.

Early studies further indicated that fatty acids were also derived from acetate in yeast[3] and the mold *Neurospora*.[4] Cell-free extracts of the bacterium *Clostridium kluyveri* could form short-chain fatty acids from ethanol *via* the intermediate formation of acetyl-CoA.[5] Similarly plant systems incorporated acetate into long-chain fatty acids and these were isolated after conversion into lipids.[6]

The involvement of C_2 units derived from acetate in the biosynthesis of the long-chain fatty acids in various mammalian tissues, microorganisms and plants was thus established. All the reactions concerned with the β-oxidation pathway whereby fatty acids are converted to acetyl-CoA in the mitochondria are reversible.[7] Suggestions concerning their biosynthesis were first advanced by Lynen and centred around the possibility that this might be accomplished by the same enzymes *via* a reversal of the β-oxidation sequence.[7,8] The process would involve head-to-tail condensations, adding

C_2 units sequentially to the carboxyl group of the growing fatty acid chain, followed by reduction, dehydration and further reduction at each stage. Indeed, some C_8 and C_{10} acids were formed from hexanoyl-CoA and [^{14}C]acetyl-CoA by mitochondrial extracts, but attempts to synthesize palmitate or stearate from acetyl-CoA, NADH, NADPH and the five enzymes of β-oxidation proved unsuccessful.[9]

It is now known that three major pathways exist for the synthesis of fatty acids.[10] The enzymes responsible in animal tissues, fungi and bacteria for *de novo* synthesis from acetyl-CoA are located in the soluble portion of the cell. This pathway utilizes NADPH as hydrogen donor for the two reduction steps at the β-oxoacyl and α,β-unsaturated acyl levels and involves the sequential addition of malonyl units, coupled with decarboxylation. All the intermediates are bound covalently to protein and the final product is usually the C_{16} or C_{18} acid. Another process is mitochondrial in origin and utilizes preformed C_{16} or C_{18} acyl-CoA derivatives (formed by the above 'malonate' pathway) with acetyl-CoA as the source of the additional C_2 units. The chain may be extended by stepwise addition to C_{22} and C_{24} acids, common components of nervous system sphingolipids. Finally, there is a similar pathway in the endoplasmic reticulum (microsomes) also engaged in lipid synthesis that requires malonyl-CoA as the condensing agent.

DE NOVO SYNTHESIS OF C_{16} AND C_{18} ACIDS

A major stimulus towards solving the problem of fatty acid synthesis occurred when malonyl-CoA was identified by Wakil[10] as the first intermediate in this pathway. Indications that fatty acid synthesis might, in fact, occur by a process which differed qualitatively from that of β-oxidation were provided by Gurin and coworkers[11] using extracts derived from the supernatant fraction of pigeon liver and later by Popják and Tietz[12] who used cell-free preparations from mammary gland. These workers obtained active extracts that catalysed the *de novo* synthesis of long-chain fatty acids, principally palmitate. (At physiological pH values, fatty acids, phosphates and other biological acids are at least partially ionized and will therefore generally be referred to throughout the text as their corresponding anions, when they are present in biological systems.) Wakil[10] subsequently demonstrated an absolute requirement for acetyl-CoA, HCO_3^-, ATP, Mn^{2+} and NADPH in avian liver systems but administration of $^{14}CO_2$ did not result in incorporation of radioactivity into fatty acids. A soluble preparation derived from plant mitochondria also required CO_2. A catalytic role for this component was therefore indicated.

Two distinct enzyme fractions from the high-speed supernatant (centrifuged at 100,000 g for 1 h) and free from particulate matter were required for maximal activity. Addition of mitochondria or microsomes had no effect on the production of fatty acids. The first fraction catalysed the carboxylation of acetyl-CoA into malonyl-CoA whereas the second enzyme fraction catalysed the conversion of acetyl-CoA and malonyl-CoA into palmitate in the presence of NADPH as hydrogen donor.[13] The free carboxyl group of malonyl-CoA (which originated from HCO_3^-) was eliminated. Lynen[14] expressed the view that malonyl-CoA might act as acceptor of the primer acetyl-CoA in the condensation reaction leading to the formation of the acetoacetyl group. The CO_2 initially incorporated into malonyl-CoA would be released at this stage.

Acetyl-CoA Carboxylase [Acetyl-CoA:Carbon Dioxide Ligase (ADP), EC 6.4.1.2]

The first intermediate in the synthesis of fatty acids by the soluble portion of the cell is malonyl-CoA[15] and this is formed by the action of acetyl-CoA carboxylase, with the concomitant hydrolysis of ATP into ADP and inorganic phosphate. Malonyl-CoA possesses a relatively strong anionic methylene group, compared with the methyl group of acetyl-CoA, that contributes towards the ease of formation of carbon–carbon bonds. The release of CO_2 helps to establish a favourable equilibrium. Wakil and Gibson[16] had previously shown that bicarbonate was required for fatty acid synthesis and that one of the enzymes possessed a high biotin content. It had been suggested much earlier that biotin might function as a coenzyme for certain carboxylation reactions. Moreover, a sparing effect of oleate and other unsaturated fatty acids on the requirement of biotin in *Lactobacillus casei* and related organisms[17] had been demonstrated since supplementation of the culture medium with oleate (or aspartate) completely removed the need for this cofactor.[18] Some relationship between biotin and fatty acid synthesis was therefore established.

The participation of biotin in the acetyl-CoA carboxylase reaction was recognized with the use of avidin as inhibitor.[15,16] Avidin is a basic protein found in raw egg-white that contains a specific binding site for biotin. It has a very high affinity for biotin and forms a stoichiometric complex which is stable but may be dissociated by heat treatment or acid hydrolysis. Avidin is now used as a diagnostic tool for the detection of biotin-dependent enzymes. The biotin is covalently bound to the apo-enzyme through an amide linkage between the ε-amino group of a lysine residue and the carboxyl group of the substituted valeric acid side-chain, as shown in (I). It can only be released by acid hydrolysis or tryptic digestion (to give biotinyl lysine). Propionyl-CoA also behaves as a good substrate for this enzyme and may be carboxylated at approximately half the rate of acetyl-CoA to form methylmalonyl-CoA.[15]

Citrate and to some extent related tricarboxylic acid cycle intermediates stimulate the activity of the carboxylase several-fold in many mammalian and avian systems. Details concerning this activation, and inhibition by

(I)

long-chain acyl-CoA derivatives will be presented at length in Chapter 4, together with a discussion of their possible physiological significance.

Excellent accounts giving full details of the early work on fatty acid biosynthesis have been presented by Lynen[19] and Wakil.[10]

Mechanism of Action of Biotin-Dependent Enzymes

Kaziro and Ochoa[20] first characterized an enzyme capable of CO_2 fixation. Propionyl-CoA carboxylase (EC 6.4.1.3), isolated from pig heart, converted propionyl-CoA into methyl malonyl-CoA. Later, other carboxylase enzymes were discovered that utilize β-methylcrotonyl-CoA, acetyl-CoA, pyruvate and geranoyl-CoA as acceptor molecules. Apart from pyruvate, the substrates concerned are coenzyme A esters and, in all cases, possess a hydrogen atom that is activated by a carbonyl group (belonging to either a thioester or ketone) that is adjacent to it or separated from it by a vinyl group.[21]

Work in Lynen's laboratory with β-methylcrotonyl-CoA carboxylase (EC 6.4.1.4) (an enzyme concerned with the catabolism of leucine) has established many of the chemical details involved in the process of CO_2 fixation by biotin-dependent enzymes.[14] These reactions are catalysed in two stages. The first is common to all and involves an ATP-phosphate exchange for which is a metal ion requirement, that results in the formation of a carboxy-biotinylenzyme complex. The reaction is driven in the forward direction by the energy released during the cleavage of the terminal phosphate of ATP. This is followed in the second stage by the transfer of CO_2 from this complex to the appropriate acceptor substrate, in this case β-methylcrotonyl-CoA, to form β-methylglutaconyl-CoA (equations (1)–(3)):

$$ATP + HCO_3^- + \text{biotinylenzyme} \rightleftharpoons \text{carboxybiotinylenzyme} + ADP + P_i \qquad (1)$$

Carboxybiotinylenzyme + $CH_3.C(CH_3):CH.CO.S.CoA \rightleftharpoons$

$$\text{biotinylenzyme} + CH_3.C(CH_2.CO_2H):CH.CO.S.CoA \qquad (2)$$

Overall: $ATP + HCO_3^- + CH_3.C(CH_3):CH.CO.S.CoA \rightleftharpoons$

$$ADP + P_i + CH_3.C(CH_2.CO_2H):CH.CO.S.CoA \quad (3)$$

The carboxybiotinylenzyme intermediate was next isolated in various laboratories; the first to be obtained was the propionyl-CoA carboxylase derivative from pig heart.[22] One mole of 'CO$_2$' was present per mole of protein-bound biotin; a stoichiometrical relationship therefore exists between the cofactor and CO$_2$.

With the aid of exchange experiments based on the use of β-methyl-crotonyl-CoA carboxylase, Lynen[14,23] also showed that free biotin, when present in high concentration and in the absence of other substrates, could replace the normal substrate and become carboxylated. More recently it has been noted[24] that a sub-unit of acetyl-CoA carboxylase from *E. coli* may also carboxylate free biotin (equation (4)):

$$ATP + HCO_3^- + \text{biotin} \rightleftharpoons ADP + P_i + \text{carboxybiotin} \quad (4)$$

Carboxybiotin is very unstable in acid conditions and has a half-life of approximately 2 hours at neutral pH at 20°. It may be readily stabilized, however, by conversion into the *N*-carboxymethylbiotin methyl ester (II) after treatment with diazomethane.[25] The product has been characterized as the 1'-*N*-isomer by comparison with authentic 1'-*N*- and 3'-*N*-derivatives.

$$CH_3.O.C-N_{1'}^{2'}{}_{3'}NH$$

(II)

Lynen predicted that the same linkage should be present in carboxybiotinyl-enzyme, an observation that has since been verified for enzyme-bound biotin in the course of CO$_2$ fixation and transcarboxylation[26] reactions. The carboxybiotinylenzyme complexes are considerably less stable than free carboxybiotin with a half-life of only 10 min under the same conditions of pH and temperature in the absence of substrate and even more unstable with substrates present.[22,26]

The identity of the derivatives,[23,27] produced by β-methylcrotonyl-CoA-, acetyl-CoA-, propionyl-CoA- and pyruvate carboxylase (EC 6.4.1.1),[28] and methylmalonyl-CoA-oxaloacetate transcarboxylase (EC 2.1.3.1)[26] (equation (5) was established after they had been separated from low-molecular

$$^-O_2C.CH_2.CO.CO_2^- + CH_3.CH_2.CO.S.CoA \xrightleftharpoons[\text{transcarboxylase}]{\text{oxaloacetate}}$$

$$CH_3.CO.CO_2^- + CH_3.\underset{\underset{CO_2^-}{|}}{CH}.CO.S.CoA \quad (5)$$

weight compounds by gel filtration chromatography on Sephadex. They were stabilized by conversion into their methyl esters which were then degraded with proteolytic enzymes. Nearly all the protein was removed by this means and the major product was carboxymethylbiotinyl lysine, in which the ε-peptide bond of this amino acid remained intact. The derivatives thus formed were purified by paper chromatography and identified as the 1'-N-isomer after comparison with the authentic derivatives. The reactive species in the propionyl-CoA carboxylase reaction is bicarbonate rather than CO_2. Experiments were conducted with $H_2^{18}O$ and $HC^{18}O_3^-$, two substrates containing a heavy isotope of oxygen, and showed that two of the ^{18}O atoms from bicarbonate were incorporated into the carboxylated product and the third into orthophosphate.[29] Since three ^{18}O atoms appear in the products, the utilization of CO_2 is excluded. Bicarbonate is also the active species for pyruvate carboxylase[30] and, by analogy, presumably participates in other biotin-dependent reactions.

The free energy change for the cleavage of the $N\text{-}CO_2^-$ bond in carboxy-biotinyl enzyme[26] has been determined at $\Delta F^{o'} = -4.7$ kcal/mole at pH 7.0 (within the range for 'high-energy' compounds). This value, however, is based on the use of free CO_2 as reactant and a somewhat lower value of -3.76, calculated for bicarbonate as substrate, may be more accurate.[21]

A probable mechanism for the formation of carboxybiotinylenzyme that involves a concerted reaction with ATP and bicarbonate[29] is presented in Scheme 2.1 (equation (6)). However, propionyl-CoA carboxylase[31] and pyruvate carboxylase[32] catalyse ATP-ADP exchange reactions in the absence of orthophosphate and a two-step mechanism has therefore been proposed[33] as a possible alternative (equations (7) and (8)). The mechanism for the transcarboxylation of acetyl-CoA is also given in equation (9) (Scheme 2.1).

CONVERSION OF ACETYL-CoA AND MALONYL-CoA INTO SATURATED FATTY ACIDS

Synthetases from yeast, avian and mammalian sources contain all the enzymes required to convert acetyl-CoA and malonyl-CoA into fatty acids, in the presence of NADPH. The major product of synthesis in animal systems is the saturated palmitate, together with smaller amounts of myristate (C_{14}) and stearate (C_{18}). The corresponding synthetases isolated from plant tissues and microorganisms yield mainly C_{18} acids.

The first clear insight into understanding the nature of the events engaged in fatty acid synthesis was provided by Lynen[19] and his colleagues as a result of their brilliant investigations over the past decade. Details concerning the intimate organization and functioning of the multienzyme complex that is responsible for this process in baker's yeast (Saccharomyces cerevisiae) have been obtained during the course of a long series of elegant experiments.

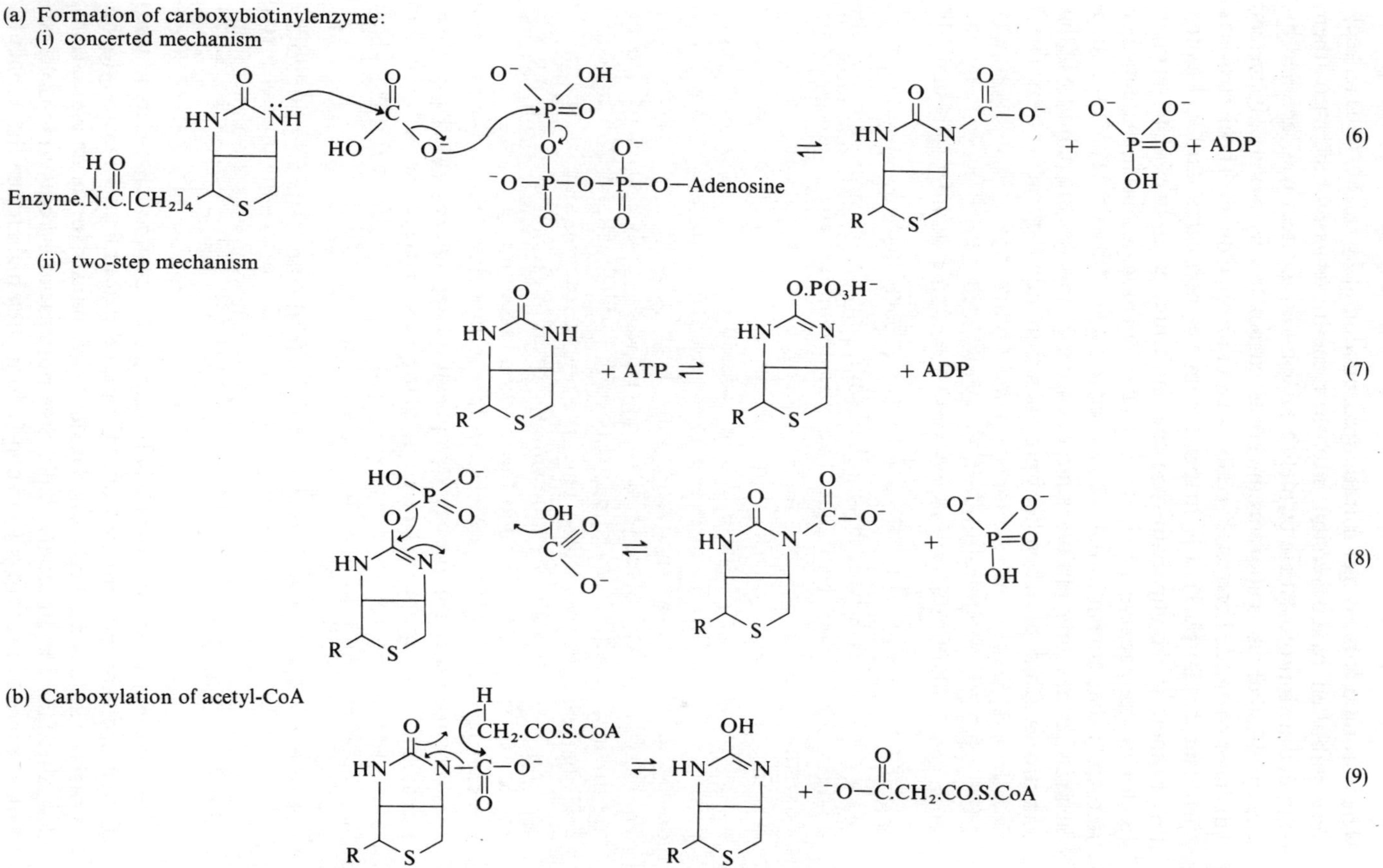

(a) Formation of carboxybiotinylenzyme:
(i) concerted mechanism

(ii) two-step mechanism

(b) Carboxylation of acetyl-CoA

Scheme 2.1 Suggested mechanisms for the ATP-dependent carboxylation of acetyl-CoA (from Lynen, 1967)[33]

The cells were broken by vigorous shaking with glass beads and eventually the purified enzyme was obtained from the cell-free extract after treatment with ammonium sulphate, $Ca_3(PO_4)_2$ gel and ultracentrifugation. This was over 200 times as active as the crude extract.[34] The enzyme was termed fatty acid synthetase and required for activity acetyl-CoA as primer, malonyl-CoA and NADPH. It was homogeneous on moving boundary electrophoresis and in the ultracentrifuge and had a molecular weight of $2 \cdot 3 \times 10^6$ [with a sedimentation coefficient $(s_{20,w})$ of 43S]. This value has been confirmed recently using a small angle X-ray scattering method.[35] The synthetase behaved as a stable multienzyme complex giving rise to palmitoyl-CoA and stearoyl-CoA ($n = 7$ or 8) as products (equation (10)):

$$\text{Acetyl-CoA} + n\,\text{malonyl-CoA} + 2n\,\text{NADPH} + 2n\,\text{H}^+ \longrightarrow$$

$$CH_3.(CH_2CH_2)_n.CO.S.CoA + n\,CO_2 + n\,CoA + 2n\,NADP^+ + n\,H_2O \quad (10)$$

It could be primed with propionyl-CoA to yield the corresponding C_{17}- and C_{19}-CoA derivatives; this mechanism is presumably responsible for the synthesis of odd-chain fatty acids. In contrast, intermediates produced during β-oxidation, e.g. β-hydroxy- and β-oxoacyl-CoA, were ineffective as precursors. Moreover, Lynen's group were unable to detect free intermediates in the course of the synthesis and this caused them to seek protein-bound products. This led to the discovery of the reaction sequences involved. Full descriptions of these experiments have been presented by Lynen.[19,33,36] The enzymic activity of the yeast synthetase was initially assayed by measuring the incorporation of radioactivity from $[2\text{-}^{14}C]$malonyl-CoA into fatty acids but with the purified enzyme, the rate of reaction may be determined spectrophotometrically by following the drop in extinction at 340 nm, due to the oxidation of NADPH as the reaction proceeds.

Experiments with $[^{14}C]$acetyl-CoA have established that, in the presence of excess malonyl-CoA, the acetyl fragment is incorporated into the terminal methyl and adjacent methylene group of the fatty acids synthesized (equation (11a)). All the other carbon atoms are derived alternately from the methylene and thioester groups from malonyl-CoA (equation (11b)):

$$^\bullet CH_3.{}^\bullet CO.S.CoA + 7\,HO_2C.CH_2.CO.S.CoA \longrightarrow {}^\bullet CH_3.{}^\bullet CH_2.[CH_2.CH_2]_6.CH_2.CO_2H$$
$$(11a)$$

$$CH_3.CO.S.CoA + 7\,HO_2C.{}^\square CH_2.{}^\circ CO.S.CoA \longrightarrow CH_3.CH_2.[{}^\square CH_2.{}^\circ CH_2]_6.{}^\square CH_2.{}^\circ CO_2H$$
$$(11b)$$

If the amount of malonyl-CoA is limiting, $[^{14}C]$acetate will actually label all the carbon atoms, since malonyl-CoA is derived directly from acetyl-CoA by carboxylation and this labelled material will be used for the condensations. The alternate carbon atoms would then possess almost equal radioactivity, with the specific activity of C-15 and C-16 slightly higher than the remainder (equation (12)):

$$\blacksquare CH_3.\bullet CO.S.CoA + 7\,HO_2C.CH_2.CO.S.CoA \rightarrow$$

$$\blacksquare CH_3.\bullet CH_2.[\blacksquare CH_2.\bullet CH_2]_6\,\blacksquare CH_2.\bullet CO_2H \quad (12)$$

The enzyme systems from animal and plant tissues, and bacteria form free acids whereas the yeast and fungal multienzyme systems give rise to the coenzyme A derivatives (equation (10)).

Condensation Reaction

Since acetyl-CoA (or a closely related homologue) is essential to initiate chain elongation, the first substantive reaction catalysed by the fatty acid synthetase is the condensation of the acyl moiety with that derived from malonyl-CoA. A short incubation of [^{14}C]acetyl-CoA with the yeast complex gave rise to radioactive material in the fraction that was precipitated with trichloroacetic acid. This stable intermediate was purified by gel filtration on Sephadex and identified as acetyl-enzyme, a high molecular weight product, in which the acetyl group was covalently linked to the protein. When this substrate was incubated with malonyl-CoA and NADPH, virtually total incorporation into fatty acids occurred. Similarly, when [1-^{14}C]acetyl-CoA was incubated with the synthetase in the presence of malonyl-CoA alone, ^{14}C-labelled acetoacetyl-enzyme was formed. Alkaline hydrolysis liberated [3-^{14}C]acetoacetate which was converted into labelled acetone after decarboxylation (equation (13)):

$$CH_3.\bullet CO.CoA + HO_2C.CH_2.CO.S.CoA \dashrightarrow CH_3.\bullet CO.CH_2.CO_2H \rightarrow$$

$$CH_3.\bullet CO.CH_3 + CO_2 \quad (13)$$

Further degradative studies[19] with ^{14}C-labelled substrates confirmed that C-1 and C-2 were derived from malonyl-CoA and C-3 and C-4 (the methyl terminal end) from acetyl-CoA.

The equilibrium constants for the reactions concerned in the synthesis of the acetoacetyl group from acetyl-CoA and malonyl-CoA (equation (14)), and the corresponding degradative action of acetoacetyl-CoA thiolase (EC 2.3.1.9) have been given by Lynen[33] (equation (15)):

$$CH_3.CO.S.CoA + HO_2C.CH_2.CO.S.CoA + HS.E + H^+ \rightleftharpoons$$

$$CH_3.CO.CH_2.CO.S.E + CO_2 + 2\,CoASH \quad (14)$$

$$CH_3.CO.S.CoA + CH_3.CO.S.CoA \rightleftharpoons CH_3.CO.CH_2.CO.S.CoA + CoASH \quad (15)$$

The value determined for K_{eq} for the reactants of equation (14), on eliminating the H^+ concentration factor was 2×10^{-2} M. The corresponding value for the β-oxothiolase reaction (expressed in the direction of synthesis of acetoacetyl-CoA) was very considerably lower at 1.6×10^{-5} M. This advantageous shift in the equilibrium position in the formation of the acetoacetyl thioester grouping is gained as a consequence of decarboxylation (ultimately due to the hydrolysis of ATP in the acetyl-CoA carboxylase reaction). Chemically the condensation reaction represents an acylation of a malonyl thioester. The nucleophilic methylene group is attracted to the electrophilic carbonyl carbon atom within the thioester group of the acetyl moiety.[19] A possible mechanism (a) for this reaction, in which displacement and decarboxylation are concerted, has been given by Lynen[33] and is illustrated in Scheme 2.2.

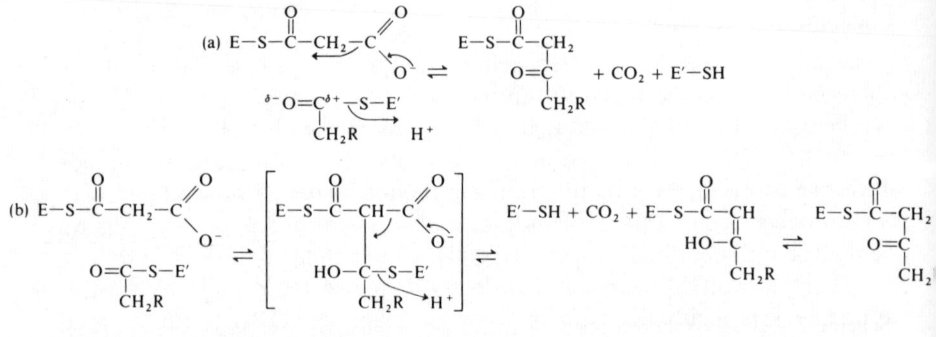

Scheme 2.2   Possible mechanisms of formation of β-oxoacyl thioesters from malonyl thioester (from Lynen, 1967)[33]

R may represent H in the case of acetyl-CoA, or alkyl groups of the type $CH_3.CH_2$ — in the case of the higher homologues. An alternate mechanism (b) was also suggested in which an intermediate carboxylated product is converted into the enol form of the β-oxoacyl derivative that finally tautomerizes to the keto form. The possible production of the enol intermediate, with the consequent potential of cis- and trans-isomerism, has interesting implications in the mode of synthesis of the numerous aromatic (single or multi-ringed) polyketides secreted by fungi. This facet will be discussed at length in Chapter 6.

Essentially a mechanism is provided in this condensation reaction by which the thermodynamically unfavourable β-oxothiolase reaction is bypassed (Scheme 2.3). The exergonic nature of the cleavage of the carboxybiotinylenzyme has already been mentioned. The ΔF value for this reaction is sufficient to enable the carboxybiotin derivative to act as a carboxylating agent for suitable acceptors.[33] The decarboxylation drives the reaction in the direction of condensation. Wood and Utter[21] point out that this situation is very similar to that encountered in the synthesis of phosphoenolpyruvate during gluconeogenesis. In this latter case, the energetically unfavourable synthesis of phosphoenolpyruvate [by the reversal of the pyruvate kinase (ATP: pyruvate phosphotransferase, EC 2.7.1.40) reaction] is circumvented by the carboxylation of pyruvate and use of an additional molecule of GTP. Further, since the enzymes concerned in the synthetic processes shown in

Scheme 2.3 are quite distinct from those involved in fatty acid oxidation or glycolysis, the possibilities of control at the enzyme level is greatly increased.

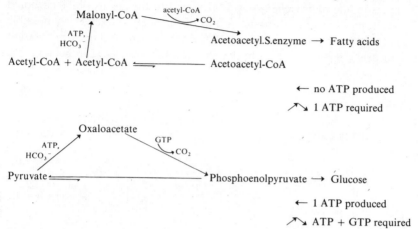

Scheme 2.3 Carboxylation reactions involved in fatty acid synthesis and gluconeo-genesis

Fatty Acid Synthetase

A knowledge of the reaction sequences involved in fatty acid synthesis was gained by Lynen with cell-free enzyme systems derived from yeast. Similar results were obtained shortly afterwards by Vagelos, Wakil, Bloch, Stumpf and coworkers with animal, bacterial and plant systems. Details appertaining to the avian and mammalian processes are very similar to that in yeast but there are sufficient differences in the bacterial and plant processes to warrant separate discussion; these will be dealt with in Chapter 3. A summary of the reactions catalysed by the yeast fatty acid synthetase has been provided by Lynen.[36] Two different categories of thiol group were identified as having a carrier function and were termed the 'central' and 'peripheral' groups in an attempt to denote their relative positions within the complex. Malonyl-CoA reacts with the central thiol group which is designated in Scheme 2.4 as the top position (HS_c—). R represents H in acetyl-CoA or an alkyl group in the butyryl- or capronyl-derivatives for instance.

The synthetase catalyses an orderly and progressive sequence of reactions which is initiated by the transfer of the acetyl group from acetyl-CoA first to the 'central' and then to the 'peripheral' thiol group. Next, a malonyl unit is similarly transferred to the 'central' thiol position. A condensation reaction occurs between the two acyl residues, with the release of CO_2, to give an acetoacetyl group covalently bound to the 'central' thiol group, itself part of the enzyme complex. The acetoacetyl and other β-oxoacyl derivatives formed from the higher homologues (C_6, C_8, etc.) are next

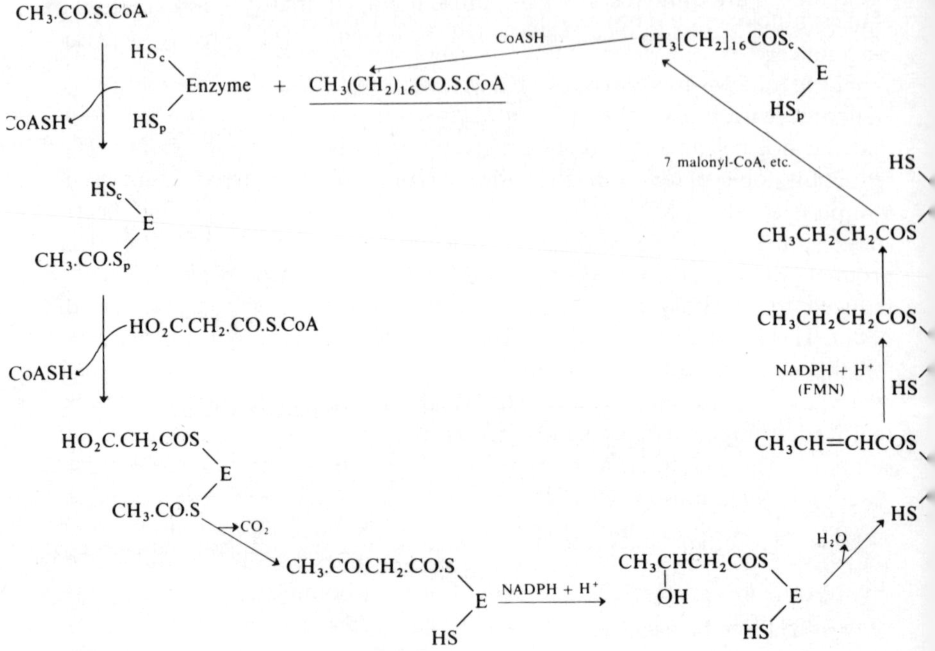

$CH_3.CO.S.CoA$

HS_c
Enzyme $+$ $CH_3(CH_2)_{16}CO.S.CoA$ CoASH $CH_3[CH_2]_{16}COS_c$ — E

CoASH HS_p HS_p

7 malonyl-CoA, etc.   HS

HS_c — E

$CH_3.CO.S_p$ $CH_3CH_2CH_2COS$ — E

$HO_2C.CH_2.CO.S.CoA$ $CH_3CH_2CH_2COS$

CoASH   NADPH $+ H^+$ (FMN)   HS

$HO_2C.CH_2COS$ — E

$CH_3.CO.S$ $CH_3CH{=}CHCOS$

$\rightarrow CO_2$ HS

       H_2O

$CH_3.CO.CH_2.CO.S$ — E $\xrightarrow{\text{NADPH} + H^+}$ $CH_3\underset{OH}{CH}CH_2COS$ — E

HS   HS

Scheme 2.4   Mechanism of fatty acid synthesis in yeast

step-wise converted into the saturated derivatives by reduction with NADPH into the D(−)-β-hydroxyacyl-enzyme, dehydration to the α,β-unsaturated acyl-enzyme and a second reduction utilizing NADPH and enzyme bound FMN as hydrogen carrier. All the intermediates involved in these reactions are bound to the 'central' thiol group. This group is also the acceptor site for malonyl units and must therefore be made available for a further round of chain-lengthening sequence. This is achieved by transfer of the butyryl- and other saturated acyl-derivatives to the 'peripheral' site, liberating the 'central' thiol site. This process is repeated six or seven times until the chain-length reaches C_{16} or C_{18}. The final reaction in the yeast synthetase is the transfer of the palmitoyl or stearoyl group from the 'central' thiol site to coenzyme A. The multienzyme complex may then react with further mole-cules of acetyl-CoA and malonyl-CoA to initiate the sequence again.

Reduction and Dehydration Reactions

Acetoacetyl-enzyme is the product of the condensation reaction and the subsequent intermediates are protein-bound in a similar manner. Lynen[19] examined the reductase and dehydratase reactions independently of each other with model compounds, since the appropriate acyl thioester derivatives of the synthetase could not readily be obtained. He used acetoacetyl-CoA and S-acetoacetyl-N-acetylcysteamine as thiol-bound substrates. The corresponding D(−)-β-hydroxybutyryl derivative was formed when they were incubated with yeast fatty acid synthetase and NADPH, but at a slow rate that did not permit the subsequent dehydration to occur appreciably. The reduction took place even after the enzyme had been freed from flavin mononucleotide (FMN). The dehydration was next shown using the D-stereoisomer of β-hydroxybutyryl-N-acetylcysteamine as substrate. The product was the crotonyl ($CH_3.CH:CH.CO-$) derivative which could be reduced to the butyryl compound by the synthetase in the presence of NADPH. This second reduction required FMN, a cofactor that is also needed for the synthesis of long-chain fatty acids in the *Mycobacterium phlei* synthetase[37] but in no other system tested. The flavin component could be removed from the yellow synthetase by treatment with acid ammonium sulphate. The resulting precipitate was colourless and inactive but was reactivated on addition of FMN.[19]

From the above account, it is evident that the basic mechanism for fatty acid synthesis is essentially the same as that for β-oxidation. The differences involved lie in such details as the use of protein-bound substrates (attached to 4′-phosphopantetheine rather than coenzyme A), production of the D-stereoisomer of β-hydroxyacyl derivatives and the site of the reaction sequence within the cell. These modifications ensure that the biosynthetic and degradative pathways remain separated.

AVIAN AND MAMMALIAN SYNTHETASE COMPLEXES

Investigations into the nature and catalytic activities of fatty acid synthetases of animal origin were pursued concurrently with the work on the yeast enzyme. In particular, early studies by Wakil and his colleagues[10,38] had shown in an elegant series of experiments that fatty acid synthetase isolated from the supernatant extract of pigeon and chicken liver catalysed the conversion of acetyl-CoA and malonyl-CoA into long-chain saturated acids in the presence of NADPH. Acetyl-CoA acted as primer and was incorporated into the terminal $CH_3.CH_2-$ group; malonyl-CoA provided the remaining C_2 units that comprised C-1 to C-14 of palmitate. Propionyl-CoA could substitute as primer with the formation of the odd-numbered C_{17} acid. Intermediates did not accumulate and a multienzyme complex again proved to be the catalytic agent.[39] The reaction mechanism therefore appeared to be similar to that described for the yeast enzyme and the overall stoichiometry for the synthesis of palmitate is as represented in equation (16):

$$CH_3.CO.S.CoA + 7\,HO_2C.CH_2.CO.S.CoA + 14\,NADPH + 14\,H^+ \rightarrow$$

$$CH_3.[CH_2]_{14}.CO_2H + 8\,CoASH + 7\,CO_2 + 14\,NADP^+ + 6\,H_2O \quad (16)$$

Purified preparations of fatty acid synthetase were also obtained from other animal sources including brain,[40] adipose tissue[41,42] and rat liver[43] and their properties were examined. In all cases the intermediates were bound to protein *via* the thiol group of 4'-phosphopantetheine, and the major product was palmitate.

Mammary gland tissue, however, tends to be rather different from other systems in that acids with shorter chain-lengths are released from the synthetase in significant amounts. The purified enzyme from lactating rat synthesized acids ranging from C_8-C_{18}.[44] Supporting experiments were performed with material from goat mammary gland[45] but the more normal pattern of $C_{14}-C_{18}$ acids was favoured when malonyl-CoA was provided in excess.[46] In addition, mammary gland tissue also possesses enzymes capable of converting acetyl-CoA into butyrate, *via* the intermediate production of acetoacetyl-CoA and its subsequent reduction with NADH.[45] This pathway is insensitive to avidin and depends upon the presence of cytoplasmic enzymes with similar properties to those engaged in β-oxidation. Indeed, recent evidence confirms the role of butyryl-CoA, formed from acetyl-CoA by this sequence, as an important primer for the synthetase in this tissue.[47]

The capacity of the mammary gland synthetases to make short- and medium-chain acids is reflected in the composition of fatty acids in milk lipids. The proportion of short-chain acids (C_4-C_{10}) is appreciable while that of C_{12} and C_{14} acids is considerable compared with tissues such as liver and adipose. Figures for the composition of milk lipids of various species are available[48] and this distribution appears to be generally valid for many mammals.[49]

Detailed studies on the enzyme from pigeon liver have been carried out by Porter and coworkers who used dithiothreitol during the isolation and purification procedures as a means of stabilizing the synthetase.[50,51] It behaved as a single homogeneous protein on sedimentation, electrophoresis and chromatography, and had a molecular weight of 4.5×10^5. This value is considerably smaller than that calculated for the yeast enzyme but approaches the value of its 'sub-unit' (based on 2.3×10^6 for the active trimer). Five N-terminal amino acids were determined for the synthetase. Electrophoresis of its components, after dissociation with urea, on polyacrylamide gel gave five to eight peptides but these did not retain any enzymic activity.[50] Thus this avian synthetase exists as a multienzyme complex in which the individual components are held together by non-covalent forces. The interactions of these enzymes are apparently vital for activity. Analysis indicated that flavin (either as FMN or FAD) was absent,[52] a notable difference from the yeast enzyme. A scheme of reactions for the production of fatty acids by the pigeon liver synthetase, very similar to that proposed earlier by Lynen (Scheme 2.4), has also been presented.[39] The sequence of reactions is repeated seven times until finally palmitate is formed by the action of a hydrolase (thioesterase) on the acyl thioester group, with liberation of the thiol group belonging to 4'-phosphopantetheine.

IDENTIFICATION OF THIOL- AND NON-THIOL ATTACHMENT SITES

The 'peripheral' thiol group as designated by Lynen belongs to a cysteine residue within the condensing enzyme component.[33] It is extremely sensitive to alkylating agents but preincubation with acetyl-CoA and, to a lesser extent saturated acyl-CoA compounds, protects it from inhibition. One mole of yeast synthetase contains three moles of 'peripheral' cysteine, probably indicating that the active multienzyme complex exists in the trimeric form.[33]

The 'central' thiol group has been identified as 4'-phosphopantetheine, bound through a phosphodiester linkage to a specific protein within the complex. A procedure for the isolation of this component from the *E. coli* synthetase had previously been described by Wakil[53] and Vagelos[54] and their colleagues. 4'-Phosphopantetheine was bound covalently to this protein by means of a phosphodiester bond to a serine residue; a more detailed description of this work will be given in Chapter 3. The same procedure was followed with the yeast enzyme.[55] It was boiled with alkali and the protein was precipitated by the addition of trichloroacetic acid. This treatment released a low molecular weight compound that was identified as 4'-phosphopantetheine after conversion into the S-benzoyl derivative with subsequent characterization by chemical analysis and comparison with authentic material.

Further support for the structure of this thiol-site in yeast has been provided by the demonstration that a pantothenate-requiring strain of baker's yeast incorporated [1-14C]pantothenate into the fatty acid synthetase with only a small loss in specific activity (on the basis of three 'subunits' per mole of active enzyme). 14C-labelled 4'-phosphopantetheine was released from the enzyme by treatment with alkali, a property which indicated that it was connected to the protein by a phosphodiester link to serine.[55,56] Conclusive evidence for its identity was obtained after enzymic conversion of this material into benzoyl-CoA. Thus the yeast enzyme complex appears to be comprised of three functional sub-units of the synthetase each containing one 'peripheral' thiol group (cysteine) and one 'central' thiol group (4'-phosphopantetheine).

Acyl Carrier Protein

More recently Lynen's group[57] has shown directly that the 4'-phosphopantetheine is actually attached to a structural protein component (Scheme 2.5) of comparatively low molecular weight within the multienzyme complex that is analogous to the acyl carrier protein of bacteria and plants

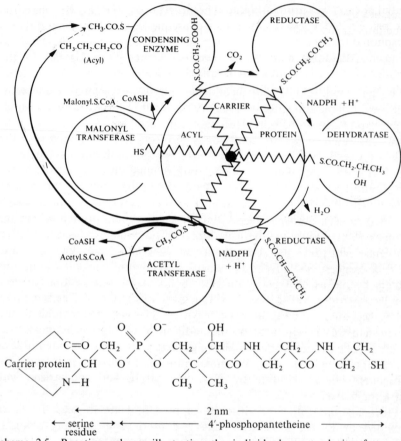

Scheme 2.5   Reaction scheme illustrating the individual events during fatty acid synthesis on a multienzyme complex (after Lynen *et al.*, 1968)[36]

(Chapter 3). ^{14}C-labelled synthetase prepared from [1-^{14}C]pantothenate was dissociated by the addition of 6M guanidine hydrochloride and the radioactive protein was resolved into various fractions by chromatography on Sephadex G-200, with the larger unlabelled proteins eluted first. The appropriate fractions were pooled and subjected to analytical electrophoresis on polyacrylamide gel. The purified material was analysed for its amino acid content and gave a molecular weight approximately 16,000, on the assumption that each ACP chain has one molecule of 4'-phosphopantetheine as prosthetic group. This value is somewhat higher than those reported[58] for bacteria and plants (9–12,000) and, in addition, there is a relatively high proline, tyrosine, phenylalanine and arginine content.[36]

It is a longstanding observation[59] that the pantothenate content of cells could not be accounted for simply by their coenzyme A content and a substantial portion was protein-bound. The structure of ACP may be considered as a 'proteinated' coenzyme A.[57] As a consequence of the binding of the 4'-phosphopantetheine prosthetic group to the carrier protein (and not to adenosine-3',5'-diphosphate as in coenzyme A), many binding sites for the enzymes of the yeast (and animal) fatty acid synthetases may be provided. In the intact complex there is firm binding of the individual enzymes and most treatments that split the multienzyme complex into sub-units cause a simultaneous loss of all enzymic activity.

The fact that malonyl and other acyl intermediates are attached to the thiol group of protein-bound 4'-phosphopantetheine underlines its particular role in fatty acid synthesis.[33,60] The sixteen bonded unit provides a flexible arm of approximately 2 nm (20 Å) length when fully extended but shorter when folded and is therefore very suitable for functioning within a multienzyme complex (Scheme 2.5). The pantetheine unit presumably rotates in such a manner that the intermediates pass in close proximity to the active sites of the individual enzymes. In this scheme, the zigzag line represents 4'-phosphopantetheine bound to the carrier protein which is itself attached in some way to the six main functioning enzymes (drawn as circles), with the gaps in the circles representing active sites. Although the scheme presented above is largely hypothetical, Lynen[33] has provided some evidence in support of the structural organization suggested in his model. Examination of electron micrographs of the purified synthetase revealed single ovoid particles with diameter 21–25 nm and a structure consisting of three interlocking subunits was apparent. Next, some confirmation that the structural unit possessed seven proteins was gained after the detection of seven N-terminal amino acids. It was proposed that these enzymes were arranged around the carrier protein that contains the 'central' thiol group (Scheme 2.5).

Information regarding the nature of the acyl-binding sites in animal systems has been obtained after incubation of [14C]acetyl-CoA and [14C]malonyl-CoA with pigeon liver synthetase.[61,62] The individual products were subjected to peptic digestion. Two [14C]-labelled peptides were isolated in each case that had the same electrophoretic mobility as peptides formed after incubation of the synthetase with [14C]pantothenate. On further hydrolysis, these peptides gave rise to acetate, malonate or acetoacetate, β-alanine and cysteamine but cysteine was absent. Accordingly, they had contained acyl residues bound to 4'-phosphopantetheine in the intact synthetase.

Porter and coworkers[63] have recently confirmed and extended these studies using techniques similar to those reported by Lynen for the yeast enzyme. The identity of 4'-phosphopantetheine was established after its conversion into the S-carboxymethylcysteamine derivative. In addition, a cysteine residue was also determined as a thiol attachment site, specific for the acetyl group, whereas a hydroxy amino acid was implicated in a third non-thiol site (see below). Treatment of the synthetase with malonyl-CoA together with hexanoyl-CoA as primer led to the formation of β-oxo-octanoyl-enzyme.[64] This C_8 product was bound by thiol linkage to 4'-phosphopantetheine and could be reduced with NADPH. The octanoyl group thus formed was bound at both the 4'-phosphopantetheine and cysteine sites. This indicated that a transferase component was active, as originally suggested by Lynen, and that the subsequent condensation reaction with a malonyl residue occurred with the saturated acyl group bound to the 'peripheral' cysteine site.

Non-Thiol Sites of Attachment

Lynen[33] first established that acetyl and malonyl groups are not, in fact, bound exclusively to the synthetase complex by means of thiol groups. Considerable progress has since been achieved in delineating the sites of attachment of acyl groups to various components of this enzyme complex.

Performic acid oxidation of thioesters converts them into their sulphonic acids with release of the acyl group as the corresponding carboxylic acid (equation (17)):

$$CH_3.CO.S.enzyme \xrightarrow{HCO.O.OH} CH_3.CO_2H + enzyme.SO_3H \qquad (17)$$

However, performic acid treatment of [^{14}C]acetyl-enzyme, prepared by reacting a constant amount of yeast enzyme with increasing concentration of [^{14}C]acetyl-CoA, demonstrated conclusively that about half the radioactivity remained protein-bound in each case and was not released as the free acid. These results and others indicated that acyl transfer to a protein component in the multienzyme complex initially occurred at non-thiol groups.[36] On this basis acetate is attached to an additional acceptor group prior to reaching the 'central' 4'-phosphopantetheine group within the ACP and subsequent transfer to the 'peripheral' cysteine group. Malonate uses a similar group before transfer to the 'central' thiol group (Scheme 2.6). Evidence for this scheme was obtained after peptic hydrolysis of the acyl-protein produced by incubation with [^{14}C]acetyl-CoA or [^{14}C]malonyl-CoA (or its methyl derivative). Chromatography of the peptic digests of (methyl)malonyl-enzyme on DEAE-Sephadex enabled three radioactive peptides to be resolved.[65] Only one reacted with performic acid and contained cysteamine and β-alanine, two constituents of 4'-phosphopantetheine. The others were stable to performic acid but sensitive to alkaline hydrolysis and were characterized as an acylheptapeptide and pentapeptide which contained serine plus six or four other amino acids respectively. The binding site therefore appears to be the hydroxyl group of a serine residue which forms an (oxygen) ester bond with the malonyl residue.

The pattern given by similar treatment of acetyl-enzyme proved to be more complex.[36] Radioactive fractions were obtained but only some were

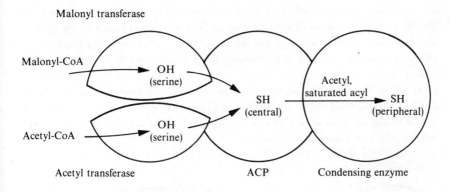

Malonyl transferase

Malonyl-CoA

OH (serine)

SH (central)

Acetyl, saturated acyl

SH (peripheral)

OH (serine)

Acetyl-CoA

Acetyl transferase   ACP   Condensing enzyme

Scheme 2.6   Specificity of acyl transfer to different acceptor groups on the multienzyme complex

present after pretreatment with iodoacetamide (which blocks the binding capacity of the 'peripheral' group). The remaining peptides contained acyl thioesters bound to cysteine residues. An acetylpantetheine-peptide was also obtained. However, a major fraction was isolated which was stable to performic acid and therefore contained an acetyl residue bound to a non-thiol site but it was quite distinct from the two serine-bound malonyl-peptides in terms of amino acid composition. It has now been characterized after tryptic digestion of [^{14}C]acetyl-peptide and contains a serine residue attached to the acetyl group.[65a] It is therefore apparent that the acetyl group is attached to this region in addition to the 'central' site at some stage, before transfer to the 'peripheral' site. The carrier groups presumably lie at the active centres of the acyl transferase enzymes (Scheme 2.6).

The order of binding to the various sites has been further examined by treating the enzyme complex with thiol reagents that exert different activities.[65] Iodoacetamide reacts preferentially with the 'peripheral' group but does not attack the 'central' group, while N-ethylmaleimide reacts with both thiol groups equally. Thus enzyme which has been treated with iodo-acetamide may accept a malonyl unit on both the 'central' 4'-phosphopante-theine and serine acceptor sites but enzyme inhibited with N-ethylmaleimide only accepts acyl groups at the serine site. Sulphur-bound radioactivity from [3-^{14}C]methylmalonyl-CoA was reduced to a very small value when the enzyme was incubated with N-ethylmaleimide. (Methylmalonyl-CoA was used in these experiments because it could transfer its acyl group to the same acceptor sites as malonyl-CoA but it did not condense with acetyl-enzyme.) In addition, transfer of methylmalonyl residues to the serine site was increased when either of the two thiol groups was blocked or after transfer of acetyl groups to the enzyme. If only the 'peripheral' site was blocked (with iodoacetamide), transfer of methylmalonate from the serine site to the 'central' thiol group was also increased. Thus there appears to be mutual interactions between the various acceptor sites. Moreover, the sequence of binding of acetyl and malonyl residues to non-thiol and thiol sites was confirmed.

Treatment of the multienzyme complex with iodoacetamide also resulted in an ability to decarboxylate the malonyl residue (Scheme 2.7) and removed the capacity for fatty acid synthesis. Acid catalysis by a group such as B^-H^+ within the condensing enzyme component is probably involved and the 'central' thiol group must remain intact for this decarboxylation.[36]

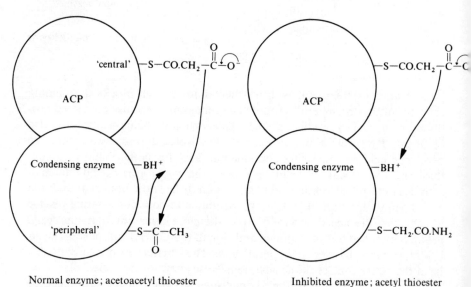

Normal enzyme; acetoacetyl thioester formation (condensation)

Inhibited enzyme; acetyl thioester formation (decarboxylation)

Scheme 2.7   Hypothetical mechanism of acetoacetate formation and decarboxylation (after Lynen et al., 1968)[36]

[^{14}C]Acyl-peptide derivatives of pigeon liver synthetase have also been examined recently after treatment with performic acid.[66,67] Serine and threonine were identified as components of these peptides that comprised the non-thiol sites. Thus both acetyl and malonyl groups are attached to the synthetase by means of an (oxygen) ester bond. Addition of iodoacetamide affected the relative affinities of acetyl and malonyl groups to thiol and non-thiol sites, causing an increase in binding to the latter.[64] This therefore appears to act as the initial binding site and is presumably a component of the transferase enzyme(s).

DISSOCIATION AND RECONSTITUTION OF FATTY ACID SYNTHETASE COMPLEXES

Yeast and animal fatty acid synthetase have been treated with reagents such as urea, guanidine hydrochloride or sodium deoxycholate, in attempts to fractionate the multienzyme complex. These methods break the non-covalent bonds that are responsible for maintaining the integrity of the synthetase and lead to disruption into smaller units but enzymic activity is usually lost irreversibly. Lynen[19] has separated the flavin component from the yeast enzyme complex and this treatment resulted in almost total loss of activity. However, addition of FMN partially restored the enoyl reductase (crotonase) and overall synthetic activity but had little effect on the other enzyme components. Enzyme activities were tested with model substrates.

Lynen's group[68] recently showed that extensive freezing of suitably buffered preparations of yeast synthetase, followed by thawing and then repeating the process, resulted in an inactive enzyme which could, however, be reactivated on dilution or dialysis. The sedimentation and electrophoretic patterns of the fatty acid synthetase had completely changed, with many subunits apparent. The number of thiol groups exposed after the enzyme complex had been dissociated greatly increased. This treatment resulted in a complete loss of fatty acid synthetase and β-oxoacyl reductase activities (the two enzymic activities tested). It was suggested that the β-oxoacyl reductase might have suffered some denaturation or possibly lacked the protein–protein interactions with neighbouring enzymes that were required for activity. On reactivation, the sedimentation and electrophoretic behaviour were identical with that of the native enzyme.

Porter and his colleagues also demonstrated in a series of experiments that the activity of the pigeon liver synthetase declined on storage especially in buffers of low ionic strength.[69] This was accompanied by dissociation of the native multienzyme complex (14S) into two inactive sub-units with a molecular weight half of that of the original protein (230,000). Incubation of the sub-units with dithiothreitol in the presence of high ionic strength buffers usually permitted reconstitution of the complex to the original molecular weight with 50 percent or greater restoration of activity. Palmitoyl-CoA also dissociated the complex into two units after breakage of non-covalent bonds in the native enzyme by mild detergent action. This treatment also severely inhibited binding of NADPH to the synthetase.[70]

Wakil and his group,[71] however, succeeded in isolating active acetyl- and malonyl-transferase components after treating the synthetase with guanidine hydrochloride, followed by partial resolution from other proteins. These enzymes also catalysed the conversion of acetyl-CoA and malonyl-CoA into their ACP (from E. coli) derivatives.

ELONGATION AND UNSATURATION MECHANISMS FOR FATTY ACID SYNTHESIS

Fatty acids that are synthesized *de novo* in animal and plant tissues consist almost entirely of the saturated palmitate and stearate (approximately 80 and 20 percent respectively). It was of natural interest to determine the location within the cell of the enzyme systems responsible for the synthesis of longer-chain unsaturated acids ($> C_{18}$) that are present as components of the various lipid classes. Attention was accordingly directed towards an understanding of the problems associated with these processes. Elongation of palmitoyl-CoA is performed by a different but mechanistically similar process in which C_2 units are added to the carboxyl end. This stepwise elongation occurs in both mitochondria and microsomes and utilizes acetyl-CoA and malonyl-CoA respectively. The lipid components of the membranous regions of these organelles are rich in these fatty acids. The first reduction and dehydration reactions leading to the synthesis of the higher homologues are mediated by β-oxoacyl-CoA dehydrogenase (L-3-hydroxyacyl-CoA:NAD oxidoreductase, EC 1.1.1.35) and enoyl hydratase (L-3-hydroxyacyl-CoA hydro-lyase, EC 4.2.1.17), both enzymes of the β-oxidation pathway. The second reduction, however, is catalysed by an independent α,β-dehydroacyl-CoA:NADPH oxidoreductase,[72] which therefore has similar cofactor requirements to that found in the synthetase. Moreover, substitution of the FAD-dependent dehydrogenase engaged in degradation with an NAD(P)H-dependent reductase would favour the synthetic sequence on thermodynamic grounds. It should also be remembered that many of these longer-chain unsaturated acids are primarily of dietary origin and are initially synthesized in plant tissues.

A procedure designed to establish the relative contributions of *de novo* synthesis or chain elongation has been developed. In general the reaction mixture is saponified with ethanolic KOH and the products subsequently acidified to liberate the fatty acids. These are extracted with solvent and the resulting extract is divided into three portions for assay of radioactivity, methylation (and subsequent analysis by gas–liquid chromatography) and decarboxylation. The ratio of the specific activity of $^{14}CO_2$ released (i.e. carboxyl carbon atom) measured as barium carbonate to that of the original fatty acid indicates the mode of synthesis. If a ratio approaching 1:1 is obtained elongation has occurred; if the ratio approaches 1:9 (for say stearate) the terminal C_2 unit has been added by the mechanism of *de novo* synthesis (Scheme 2.8). A value lying between these figures indicates that the acid has been formed to some extent by each process.

Mitochondrial System

It is difficult to explain how the initial condensation reaction is achieved in mitochondria since malonyl-CoA is not involved to overcome the

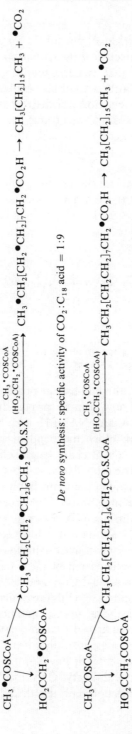

Scheme 2.8 Patterns of incorporation of acetyl-CoA and malonyl-CoA for *de novo* synthesis and elongation mechanisms

unfavourable equilibrium. However, the enzyme concerned in this condensation may be dependent on pyridoxal phosphate for activity;[10] this requirement may therefore be connected with activation of acetyl-CoA *via* the formation of a Schiff's base[73] (Scheme 2.9), that is, giving rise to a situation similar to that after carboxylation. A simpler explanation may rest on the rapid coupling of the β-oxoacyl-CoA formed by the condensing enzyme (thiolase; acyl-CoA : acetyl-CoA C-acyltransferase, EC 2.3.1.16) to the reductase that favours synthesis.[72]

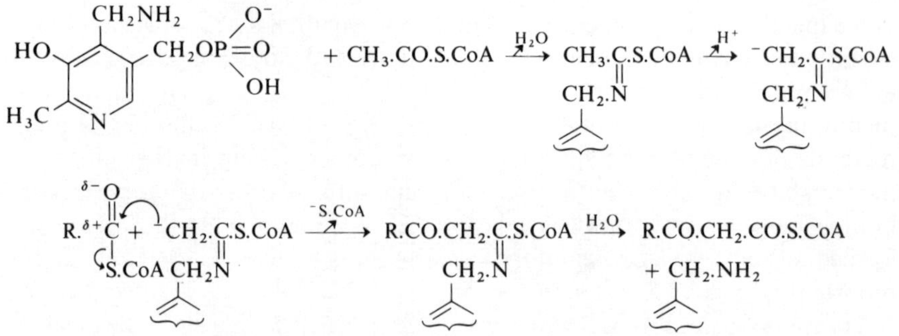

Scheme 2.9   Possible reaction mechanism for the pyridoxamine phosphate-dependent long-chain acyl-CoA condensing enzyme (from Goodwin and Mercer, 1970)[73]

Wakil's group first studied the elongation process in soluble extracts of rat liver mitochondria.[74] An acyl primer (principally palmitoyl-CoA) was needed together with acetyl-CoA, NADPH and NADH. Synthesis was insensitive to avidin and therefore independent of malonyl-CoA. Both saturated and unsaturated acids of chain-length C_{18}–C_{22}, with up to six double bonds, were formed. The type of acid synthesized agreed closely with the lipid composition of mitochondria and probably reflects the nature of the phospholipids present in the membrane. *cis*-Vaccenate ($C_{18:1}$), a normal product of *de novo* synthesis in bacteria, was also present in this fraction. Experiments in which [1-^{14}C]palmitate was injected to rats indicated that *cis*-vaccenate was formed by elongation of palmitoleoyl-CoA; radioactivity was also found in oleate. Confirmation of these results were obtained by direction conversion of palmitoleoyl-CoA and [^{14}C]acetyl-CoA into *cis*-vaccenate by rat liver mitochondria.[75] Thus introduction of the double bond for the synthesis of this acid occurs at the C_{16} and not the C_{18} stage.

Many other workers later confirmed that this pathway was operative in mitochondria from various tissues, including heart,[76] brain[77] and liver[78] from a number of species. In all cases an acyl-CoA substrate was required to prime the synthesis, together with acetyl-CoA and NADH. The nature of the fatty acids produced in respect of chain-length and degree of unsaturation was determined after examination of their methyl esters by gas–liquid chromatography before and after hydrogenation.

Submitochondrial particles comprising of inner and outer membranes and also soluble fractions have been studied to determine their capacity for elongation.[78,79] These fractions were characterized by the presence of specific marker enzymes (and their total activities) and also after electron microscopy. Unfortunately, the plethora of results concerning the actual site of elongation within the mitochondria has given rise to conflicting interpretations and the problem of the true localization(s) has not yet been resolved. Reports in the Literature concerning the ability of the inner and outer membrane fractions to perform *de novo* synthesis or elongation have recently been extended, for instance, in studies with sonicated extracts from guinea pig[80] and rat liver[81] preparations. Possibly the capacity of various membrane fractions varies from one species or tissue to another. Again, methods of assigning the sub-organelle fractions by using marker enzymes may not have been sufficiently discriminating, with the result that these lacked homogeneity; variations in degradation techniques applied to the fatty acids formed may also be a contributory factor in the discrepancies that have been reported.

Transfer of palmitate and stearate from their site of synthesis in the cytosol through the inner membrane of the mitochondria occurs *via* the intermediacy of their CoA and carnitine derivatives. The CoA substrate is regenerated in the mitochondria.[82]

Microsomal System

Mammalian microsomes (that is, particles that sediment on centrifugation at $100,000\,g$ for 1 hour after prior removal of mitochondria and nuclei) also possess the ability to elongate saturated and unsaturated acyl-CoA thioesters and convert the products into lipids.[83–85] They do not synthesize fatty acids *de novo*. Malonyl-CoA is the source of the C_2 units and the hydrogen donor is NADPH; thus a simple reversal of β-oxidation cannot be operative. The normal products are the C_{18} and C_{20} acids with a large proportion possessing unsaturation.[84] Chain elongation reactions may be examined separately from those concerned with desaturation by using anaerobic systems.[86]

Pigeon and rat liver microsomes were used in the earlier studies by a number of groups but malonyl-CoA has now been confirmed as the true precursor of the additional C_2 units in stearate and the polyunsaturated C_{22} acids in a microsomal preparation from rat brain.[77]

Synthesis of Unsaturated Acids

The end-products of the fatty acid synthetase sequence of reactions are almost invariably (with the exception of bacteria) the saturated acids, principally palmitate. Schoenheimer and Rittenberg[87] first demonstrated the direct conversion of stearate into oleate in animal tissues using the deuterated substrate. This early work was followed with the discovery that palmitate might be similarly desaturated to palmitoleate.[88] Both products possess unsaturation at C-9.

Desaturation in Yeast

A considerable interval elapsed before more definitive work was performed when Bloomfield and Bloch[89] originally implicated an aerobic mechanism for this process. They observed that anaerobically maintained yeast cells (*S. cerevisiae*) synthesized only saturated fatty acids, whereas unsaturated acids were also formed in the presence of air. Moreover, this yeast required oleate as a growth factor under anaerobic conditions. These studies gave little insight into the actual mechanism involved but were then extended to cell-free systems that could convert palmitoyl-CoA and stearoyl-CoA into their corresponding monoenoic acid. The enzyme responsible was particulate and required NAD(P)H and molecular oxygen. The intracellular site for these reactions has been located in the membranous portion of the micro-somes as, indeed, it has for many of the reactions involved in the biosynthesis of various lipids.

Another yeast, *Torulopsis utilis*, converted oleate into linoleate ($C_{18:2,\Delta^{9,12}}$) in high yield by a similar oxygen-dependent process.[90] Ricinoleate (12-hydroxyhexadec-9-enoate) did not act as a precursor in this conversion, indicating that a monooxygenase type of reaction (whereby a hydroxy acid may be formed as an intermediate and subsequently dehydrated) was not involved. Similarly, a hydroxylation–dehydration sequence is not implicated in the formation of oleate. Presumably the hydrogen atoms concerned are removed directly by the desaturase system and transferred to a specific electron transporting system as with the liver enzyme. This will now be discussed.

Desaturation in Animals

Marsh and James[91] have shown that rat liver microsomes convert stearate into oleate by a process that is similar to the yeast system. It has already been mentioned[75] that palmitate may be desaturated to *cis*-palmitoleate ($C_{16:1,\Delta^9}$) which may then be elongated to *cis*-vaccenate ($C_{18:1,\Delta^{11}}$) or alternatively may be elongated prior to desaturation. In contrast, stearate is desaturated directly to give oleate ($C_{18:1,\Delta^9}$). The biosynthetic relationship between these acids is illustrated in Scheme 2.10.

Wakil and associates[92] later purified stearoyl-CoA desaturase from hen liver microsomes and observed that NADH was the preferred electron donor. Desaturase activity was lost after acetone extraction but this was restored on addition of various lipids. It had previously been shown[93]

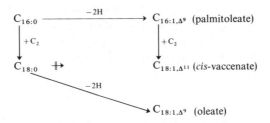

Scheme 2.10 Pattern of desaturation and elongation with $C_{16:0}$ and $C_{18:0}$ substrates

that the microsomal NADH-cytochrome c reductase was tightly attached to the membranous components and required phospholipid for activity unlike its NADPH-mediated counterpart (EC 1.6.2.3). However, it is probable that much of the lipid material is superfluous for desaturase activity.[94] The product of desaturation was characterized as oleate but this was presumably formed after enzymic hydrolysis of its coenzyme A derivative by a hydrolase contaminant.[92] Similarly, palmitoyl-CoA was desaturated to cis-hexadecenoate (palmitoleate) with unsaturation at the C-9 position. Studies implicating NADH and a cyanide-sensitive factor with cytochrome b_5 as electron carrier have also been reported by other workers.[95]

More recently, the enzyme NADH-cytochrome b_5 reductase has been identified as an integral component of a solubilized desaturase system.[96] Thus current assessment of our understanding of the desaturation process indicates that there is no direct intervention of oxygen in terms of formation of an oxygenated acyl intermediate and that the electron flow moves in the direction of NADH to cytochrome c via the flavin of the reductase and haem of cytochrome b_5. The overall composition of this electron transport chain is similar to the microsomal NADPH-cytochrome P_{450} system which is concerned with hydroxylation reactions (monooxygenase activity) but does not possess a desaturase function. A similar multicomponent preparation has been isolated from the phytoflagellate (plant) Euglena gracilis that acts upon stearoyl-ACP or stearoyl-CoA and consists of the enzymes NADPH oxidase and desaturase with ferredoxin (Chapter 3).[97]

Comparative results on the specificities of desaturases from various types of tissue have been obtained by James and colleagues for animal microsomal preparations and whole cells of Torulopsis yeast and the alga Chlorella vulgaris.[98] These systems all possess a desaturase that introduces a double bond at the C-9 position, with maximum efficiency for the C_{18} acid as substrate. Methyl substitution on any carbon atom reduced or even abolished $(C_5–C_{15})$ desaturase activity but did not affect activation to the coenzyme A thioester or incorporation into lipids.

Higher animals are unable to convert oleate into the dienoic linoleate $(C_{18:2,\Delta^{9,12}})$ by the specific mechanism of Δ^{12}-unsaturation that is operative in yeasts[99] (and plants) and hence this and related 'essential' fatty acids have to be supplied in the diet. Animals can, however, synthesize certain C_{20}–C_{24} polyunsaturated acids by a process of chain-elongation coupled with desaturation.[100,101] Coenzyme A derivatives of dietary unsaturated acids act as substrates; linoleoyl-CoA (III), for example, may be converted into the trienoic γ-linolenyl-CoA (IV) in the first instance (equation (18)):

$$CH_3.[CH_2]_4.CH:^{12}CH.CH_2.CH:^9CH.[CH_2]_7.CO.S.CoA + NADH + H^+ + O_2 \rightarrow$$

$$(III)$$

$$CH_3.[CH_2]_4.CH:^{12}CH.CH_2.CH:^9CH.CH_2.CH:^6CH.[CH_2]_4.CO.S.CoA + NAD + 2\,H_2O$$

$$(IV) \tag{18}$$

The introduction of double bonds in these animal systems is directed towards the carboxyl group in a divinylmethane pattern to form a distinct family of acids. This unsaturation is generally associated with elongation to the di-, tri- and tetra-unsaturated eicosenoates (C_{20}). The polyunsaturation system of rat liver is present in the microsomes and also requires oxygen and a reduced nicotinamide nucleotide.

Certain unsaturated acids may also be formed, often in the *trans* configuration, in the rumen of ruminant animals through partial hydrogenation of more highly unsaturated precursors (provided in the diet) by the microorganisms present.

Finally, the essential polyunsaturated acids play a further important (and possibly their most vital) role in metabolism by behaving as precursors of the prostaglandins[102] as exemplified by the conversion of arachidonic acid ($C_{20:4,\Delta^{5,8,11,14}}$) (V) into the prostaglandin PGE_2 (VI). Similar structurally-related products are formed from the eicosatrienoic ($C_{20:3,\Delta^{8,11,14}}$) and

(V)   (VI)

pentaenoic acids. The additional oxygen atoms all derive from molecular O_2. Prostaglandins are synthesized by the microsomal fraction of seminal vesicles and several other tissues. They exert potent physiological activity at very low concentration and possess vasodepressor and smooth muscle stimulating properties. At the biochemical level they antagonize the action of adrenaline and glucagon, for instance, on adipose tissue lipase; this and similar effects appear to be mediated through their action on adenyl cyclase. A recent discussion of their structure and mode of biosynthesis has been presented by Samuelsson.[102]

REFERENCES

1. Morris, L. J., *J. Lipid Res.*, **7,** 717 (1966)
2. Rittenberg, D., and Bloch, K., *J. Biol. Chem.*, **160,** 417 (1945)
3. White, A. G. C., and Werkman, C. H., *Arch. Biochem.*, **13,** 27 (1947)
4. Ottke, R. C., Tatum, E. L., Zabin, I., and Bloch, K., *J. Biol. Chem.*, **189,** 429 (1951)
5. Stadtman, E. R., and Barker, H. A., *J. Biol. Chem.*, **180,** 1085 (1949)
6. Stumpf, P. K., and Barber, G. A., *J. Biol. Chem.*, **227,** 407 (1957)
7. Lynen, F., *Angew. Chem.*, **53,** 490 (1951)
8. Lynen, F., and Ochoa, S., *Biochim. Biophys. Acta*, **12,** 299 (1953)
9. Seubert, W., Greull, G., and Lynen, F., *Angew. Chem.*, **69,** 359 (1957)
10. Wakil, S. J., *J. Lipid Res.*, **2,** 1 (1961)
11. Brady, R. O., and Gurin, S., *J. Biol. Chem.*, **199,** 421 (1952); van Baalen, J., and Gurin, S. *J. Biol. Chem.*, **205,** 303 (1953)
12. Popják, G., and Tietz, A., *Biochem. J.*, **60,** 147, 155 (1955)
13. Ganguly, J., *Biochim. Biophys. Acta*, **40,** 110 (1960)
14. Lynen, F., *J. Cell. Comp. Physiol.*, **54,** Suppl. 1, 35 (1959)
15. Waite, M., and Wakil, S. J., *J. Biol. Chem.*, **237,** 2750 (1962); *J. Biol. Chem.*, **238,** 77, 81 (1963)
16. Wakil, S. J., and Gibson, D. M., *Biochim. Biophys. Acta*, **41,** 122 (1960)
17. Williams, V. R., and Fieger, E. A., *J. Biol. Chem.*, **166,** 335 (1946)
18. Williams, W. L., Broquist, H. P., and Snell, E. E., *J. Biol. Chem.*, **170,** 619 (1947)
19. Lynen, F., *Fed. Proc.*, **20,** 941 (1961)
20. Kaziro, Y., and Ochoa, S., *Advan. Enzymol.*, **26,** 283 (1964)
21. Wood, H. G., and Utter, M. F. In *Essays in Biochemistry*, Vol. 1, p. 1. Ed. by Campbell, P. N., and Greville, G. D., Academic Press Inc., London and New York, 1965
22. Kaziro, Y., and Ochoa, S., *J. Biol. Chem.*, **236,** 3131 (1961); Ochoa, S., and Kaziro, Y., *Fed. Proc.*, **20,** 982 (1961)
23. Lynen, F., Knappe, J., Lorch, E., Jütting, G., Ringelmann, E., and Lachance, J. P., *Biochem. Z.*, **335,** 123 (1961)
24. Alberts, A. W., Nervi, A. M., and Vagelos, P. R., *Proc. Nat. Acad. Sci. U.S.*, **63,** 1319 (1969)
25. Knappe, J., Ringelmann, E., and Lynen, F., *Biochem. Z.*, **335,** 168 (1961)
26. Wood, H. G., Lochmüller, H., Riepertinger, C., and Lynen, F., *Biochem. Z.* **337,** 247 (1963)
27. Knappe, J., Wenger, B., and Wiegand, U., *Biochem. Z.*, **337,** 232 (1963)
28. Scrutton, M. C., Keech, D. B., and Utter, M. F., *J. Biol. Chem.*, **240,** 574 (1965)
29. Kaziro, Y., Hass, L. F., Boyer, P. D., and Ochoa, S., *J. Biol. Chem.*, **237,** 1460 (1962)
30. Cooper, T. G., Tchen, T. T., Wood, H. G., and Benedict, C. R., *J. Biol. Chem.*, **243,** 3857 (1968)
31. Lane, M. D., Halenz, D. R., Kosow, D. P., and Hegre, C. S., *J. Biol. Chem.*, **235,** 3082 (1960)
32. Scrutton, M. C., and Utter, M. F., *J. Biol. Chem.*, **240,** 3714 (1965)
33. Lynen, F., *Biochem. J.*, **102,** 381 (1967)
34. Lynen, F., *Methods Enzymol.*, **14,** 17 (1969)
35. Pilz, I., Herbst, M., Kratky, O., Oesterhelt, D., and Lynen, F. *Eur. J. Biochem.*, **13,** 55 (1970)

36. Lynen, F., Oesterhelt, D., Schweizer, E., and Willecke, K. In *Cellular Compartmentalization and Control of Fatty Acid Metabolism*, pl. Ed. by Gran, F. C., Universitetsforlaget., Academic Press Inc., London and New York, 1968
37. Ilton, M., Jevans, A. W., McCarthy, E. D., Vance, D., White, H. B., and Bloch, K., *Proc. Nat. Acad. Sci. U.S.*, **68,** 87 (1971)
38. Bressler, R., and Wakil, S. J., *J. Biol. Chem.*, **236,** 1643 (1961)
39. Hsu, R. Y., Wasson, G., and Porter, J. W., *J. Biol. Chem.*, **240,** 3736 (1965)
40. Brady, R. O., *J. Biol. Chem.*, **235,** 3099 (1960)
41. Martin, D. B., Horning, M. G., and Vagelos, P. R., *J. Biol. Chem.*, **236,** 663 (1961)
42. Horning, M. G., Martin, D. B., Karmen, A., and Vagelos, P. R., *J. Biol. Chem.*, **236,** 669 (1961)
43. Burton, D. N., Haavik, A. G., and Porter, J. W., *Arch. Biochem. Biophys.*, **126,** 141 (1968)
44. Dils, R., and Popják, G., *Biochem. J.*, **83,** 41 (1962)
45. Becker, M. E., and Kumar, S., *Biochemistry*, **4,** 1839 (1965)
46. Smith, S., and Dils, R., *Biochim. Biophys. Acta*, **116,** 23 (1966)
47. Smith, S., and Abraham, S., *J. Biol. Chem.*, **246,** 2537 (1971); Lin, C. Y. and Kumar, S., *J. Biol. Chem.*, **246,** 3284 (1971)
48. Garton, G. A., *J. Lipid Res.*, **4,** 237 (1963)
49. Smith, S., Watts, R., and Dils, R., *J. Lipid Res.*, **9,** 52 (1968)
50. Yang, P. C., Butterworth, P. H. W., Bock, R. M., and Porter, J. W., *J. Biol. Chem.*, **242,** 3501 (1967)
51. Butterworth, P. H. W., Yang, P. C., Bock, R. M., and Porter, J. W., *J. Biol. Chem.*, **242,** 3508 (1967)
52. Chesterton, C. J., Butterworth, P. H. W., Abramovitz, A. S., Jacob, E. J., and Porter, J. W., *Arch. Biochem. Biophys.*, **124,** 386 (1968)
53. Sauer, F., Pugh, E. L., Wakil, S. J., Delaney, R., and Hill, R. L., *Proc. Nat. Acad. Sci. U.S.*, **52,** 1360 (1964)
54. Majerus, P. W., Alberts, A. W., and Vagelos, P. R., *J. Biol. Chem.*, **240,** 4723 (1965); Vagelos, P. R., Majerus, P. W., Alberts, A. W., Larrabee, A. R., and Ailhaud, G. P., *Fed. Proc.*, **25,** 1485 (1966)
55. Wells, W. W., Schultz, J., and Lynen, F., *Proc. Nat. Acad. Sci. U.S.*, **56,** 633 (1966)
56. Wells, W. W., Schultz, J., and Lynen, F., *Biochem. Z.*, **346,** 474 (1967)
57. Willecke, K., Ritter, E., and Lynen, F., *Eur. J. Biochem.*, **8,** 503 (1969)
58. Simoni, R. D., Criddle, R. S., and Stumpf, P. K., *J. Biol. Chem.*, **242,** 573 (1967)
59. György, P. In *The Vitamins*, Vol. 2, p. 589. Ed. by Sebrell, W. H. and Harris, R. S. Academic Press Inc., New York, 1954
60. Lipmann, F., *Science*, **173,** 875 (1971)
61. Jacob, E. J., Butterworth, P. H. W., and Porter, J. W., *Arch. Biochem. Biophys.*, **124,** 392 (1968)
62. Chesterton, C. J., Butterworth, P. H. W., and Porter, J. W., *Arch. Biochem. Biophys.*, **126,** 864 (1968)
63. Phillips, G. T., Nixon, J. E., Abramovitz, A. S., and Porter, J. W., *Arch. Biochem. Biophys.*, **138,** 357 (1970)
64. Nixon, J. E., Phillips, G. T., Abramovitz, A. S., and Porter, J. W., *Arch. Biochem. Biophys.*, **138,** 372 (1970)
65. Schweizer, E., Piccinini, F., Duba, C., Günther, S., Ritter, E., and Lynen, F. *Eur. J. Biochem.*, **15,** 483 (1970)
65a. Lynen, F., *Biochem. J.*, **128,** 1P (1972)
66. Phillips, G. T., Nixon, J. E., Dorsey, J. A., Butterworth, P. H. W., Chesterton, C. J., and Porter, J. W., *Arch. Biochem. Biophys.*, **138,** 380 (1970)

67. Joshi, V. C., Plate, C. A., and Wakil, S. J., *J. Biol. Chem.*, **245**, 2857 (1970)
68. Sumper, M., Riepertinger, C., and Lynen, F., *FEBS Lett.*, **5**, 45 (1969)
69. Kumar, S., Dorsey, J. K., and Porter, J. W., *Biochem. Biophys. Res. Commun.*, **40**, 825 (1970)
70. Dugan, R. E., and Porter, J. W., *J. Biol. Chem.*, **245**, 2051 (1970)
71. Plate, C. A., Joshi, V. C., and Wakil, S. J., *J. Biol. Chem.*, **245**, 2868 (1970)
72. Seubert, W., Lamberts, I., Kramer, R., and Ohly, B., *Biochim. Biophys. Acta*, **164**, 498 (1968)
73. Goodwin, T. W., and Mercer, E. I. In *Introduction to Plant Biochemistry*, p. 178. Pergamon Press Ltd., London, 1972
74. Harlan, W. R., and Wakil, S. J., *J. Biol. Chem.*, **238**, 3216 (1963)
75. Holloway, P. W., and Wakil, S. J., *J. Biol. Chem.*, **239**, 2489 (1964)
76. Dahlen, J. A., and Porter, J. W., *Arch. Biochem. Biophys.*, **127**, 207 (1968)
77. Aeberhard, E., and Menkes, J. H., *J. Biol. Chem.*, **243**, 3834 (1968)
78. Colli, W., Hinkle, P. C., and Pullman, M. E., *J. Biol. Chem.*, **244**, 6432 (1969)
79. Whereat, A. F., Orishimo, M. W., Nelson, J., and Phillips, S. J., *J. Biol. Chem.*, **244**, 6498 (1969)
80. Wit-Peeters, E. M., *Biochim. Biophys. Acta*, **176**, 453 (1969)
81. Howard, C. F., *J. Biol. Chem.*, **245**, 462 (1970)
82. Greville, G. D., and Tubbs, P. K. In *Essays in Biochemistry*, Vol. 4, p. 155. Ed. by Campbell, P. N., and Greville, G. D. Academic Press Inc., London and New York, 1968
83. Lorch, E., Abraham, S., and Chaikoff, I. L., *Biochim. Biophys. Acta*, **70**, 627 (1963)
84. Nugteren, D. H., *Biochim. Biophys. Acta*, **106**, 280 (1965)
85. Guchhait, R. B., Putz, G. R., and Porter, J. W., *Arch. Biochem. Biophys.*, **117**, 541 (1966)
86. Mohrhauer, H., Christiansen, K., Gan, M. V., Deubig, M., and Holman, R. T., *J. Biol. Chem.*, **242**, 4507 (1967)
87. Schoenheimer, R., and Rittenberg, D., *J. Biol. Chem.*, **113**, 505 (1936)
88. Stetten, D., and Schoenheimer, R., *J. Biol. Chem.* **133**, 329 (1940)
89. Bloomfield, D. K., and Bloch, K., *J. Biol. Chem.*, **235**, 337 (1960)
90. Yuan, C., and Bloch, K., *J. Biol. Chem.*, **236**, 1277 (1961)
91. Marsh, J. B., and James, A. T., *Biochim. Biophys. Acta*, **60**, 320 (1962)
92. Jones, P. D., Holloway, P. W., Pelluffo, R. O., and Wakil, S. J., *J. Biol. Chem.*, **244**, 744 (1969)
93. Jones, P. D., and Wakil, S. J., *J. Biol. Chem.*, **242**, 5267 (1967)
94. Gurr, M. I., and Robinson, M. P., *Eur. J. Biochem.*, **15**, 335 (1970)
95. Oshino, N., Imai, Y., and Sato, R., *Biochim. Biophys. Acta*, **128**, 13 (1966)
96. Holloway, P. W., and Wakil, S. J., *J. Biol. Chem.*, **245**, 1862 (1970)
97. Nagai, J., and Bloch, K., *J. Biol. Chem.*, **243**, 4626 (1968)
98. Brett, D., Howling, D., Morris, L. J., and James, A. T., *Arch. Biochem. Biophys.*, **143**, 535 (1971)
99. Baker, N., and Lynen, F., *Eur. J. Biochem.*, **19**, 200 (1971)
100. Stoffel, W., *Biochem. Biophys. Res. Commun.*, **6**, 270 (1961)
101. Nugteren, D. H., *Biochim. Biophys. Acta*, **60**, 656 (1962)
102. Samuelsson, B. In *Lipid Metabolism*, p. 107. Ed. by Wakil, S. J., Academic Press Inc., New York and London, 1970

CHAPTER 3

Biosynthesis of Fatty Acids in Bacteria and Plants

INTRODUCTION

The fatty acid synthesizing systems of yeast and animal tissues that have been described in the previous chapter are all tightly bound multienzyme complexes that behave functionally as single units. The intermediates are bound covalently to the synthetase but the component enzymes and carrier protein cannot usually be resolved from each other without total loss of activity. A different class of synthetase occurs in most bacteria and plants. Vagelos, Bloch and Wakil with their respective groups have made especially valuable contributions in this field and have established the sequence of reactions by which fatty acids are synthesized in bacteria. *Escherichia coli* has been the principal organism used for the attainment of critical information; its synthetase readily dissociates into the individual enzymes, all of which may be fractionated and purified by conventional methods of protein separation. Stumpf and coworkers have characterized the system that occurs in certain plant tissues.

A further feature which differentiates the bacterial synthetase rests in its ability to form a C_{18} monounsaturated acid by a mechanism that is intimately linked to the other reactions involved. The process is therefore entirely independent of oxygen but is very frequently utilized even when organisms are grown aerobically. A novel aspect of bacterial lipids is that they contain fatty acids that possess *C*-methyl or biosynthetically related groups. They are derived from unsaturated acids and appear to exert similar physical behaviour. The effect in all cases is to reduce the melting point compared with the analogous saturated acid. Polyunsaturated acids and, with some minor exceptions, dienoic acids are not found in bacterial lipids. These acids with chain-length C_{18} and C_{20} are extremely common, however, in plant tissues especially seed oils and chloroplasts, and are formed aerobically. Acetylenic acids are also found in these oils but aspects of their biosynthesis will be dealt with later in Chapter 6.

Vagelos[1] and Bloch[2] and their colleagues first prepared cell-free extracts from *E. coli* that yielded fatty acid synthetic activity and converted acetyl-CoA and malonyl-CoA into palmitate and *cis*-vaccenate ($C_{18:1}$). A similar preparation from *Clostridium kluyveri* formed predominantly saturated acids.[1] Two protein fractions were isolated from these soluble extracts, one heat-labile and the other heat-stable. These studies were extended with sonicated extracts of E. coli and showed that the heat-stable protein acted as a coenzyme.[3] It possessed a terminal thiol group to which the intermediates in fatty acid synthesis were linked as acyl thioesters.[4] It has therefore been termed acyl carrier protein (ACP).

Incubation of this protein cofactor with [14C]acetyl-CoA or [14C]malonyl-CoA gave radioactive products bound to protein that were isolated after chromatography on Sephadex. This material was stable to acid but released radioactive acetate or malonate after alkaline hydrolysis. Incubation of [14C]acetyl-ACP with malonyl-CoA, NADPH, ACP and the enzyme preparation gave radioactive long-chain fatty acids. Similar results were obtained with [14C]malonyl-ACP and the appropriate cofactors. Moreover, incubation of [14C]malonyl-CoA with acetyl-CoA and ACP together with the condensing enzyme gave rise to 14C-labelled acetoacetyl-ACP. This product was then readily converted into palmitate and cis-vaccenate in the presence of NADPH, ACP and enzyme.[1,5]

ACYL CARRIER PROTEIN (ACP)

Detailed structural studies on ACP have been performed on material obtained from E. coli. It is heat- and acid-stable with molecular weight approximately 9,000, based on sedimentation and diffusion studies and also amino acid analysis.[6] One thiol group per mole of ACP is present which has been identified as a constituent of 4'-phosphopantetheine (I).[7] All the re-

$$
\begin{array}{ccccc}
 & O & CH_3 & OH & O & O \\
 & \| & | & | & \| & \| \\
HO.P.O.CH_2.C & . & CH.C.NH.CH_2.CH_2.C.NH.CH_2.CH_2.SH \\
 & | & | \\
 & O^- & CH_3 \\
\end{array}
$$

(I)

actions of fatty acid synthesis occur with the acyl substrates bound by thioester linkage to ACP. In each case, the acyl-ACP derivatives are either considerably more reactive than the corresponding acyl-CoA compounds with the individual enzymes engaged in the process or the latter may even be completely inactive.

Vagelos and Wakil and their coworkers have established that 4'-phosphopantetheine is covalently linked to the ACP apoprotein through a phosphodiester bond to a serine residue.[7,8] Incubation of [14C]malonyl-CoA or [14C]acetyl-CoA with ACP and the appropriate acyl transferase gave a single radioactive peptide that was removed after precipitation with trichloroacetic acid.[9] The malonyl- and acetyl-ACP were digested with pepsin and two radioactive peptides thus produced were used for structural studies after purification.[7-9] One of these contained Asp, Ser, Leu and 4'-phosphopantetheine (II) whereas the other overlapped and contained in addition

Gly and Ala. These peptides possessed stoichiometric amounts of the acyl group, phosphate, β-alanine, thioethanolamine and pantoic acid but adenine and ribose were absent. A coenzyme A-like product was therefore excluded. The prosthetic group was liberated from the serine residue by heating with alkali, confirming the lability of the phosphodiester bond, and was then identified categorically as 4'-phosphopantetheine after enzymic conversion into coenzyme A. Further studies on the mechanism for the alkaline hydrolysis indicated that it proceeded *via* a β-elimination that converted the serine into dehydro-alanine. This, in turn, gave pyruvic acid on acid hydrolysis (equations (1) and (2)) and alanine after catalytic reduction.

$$
\begin{array}{ccc}
\text{Asp} & & \text{Asp} \\
| & & | \\
\text{NH} \quad \text{O} & & \text{NH} \\
| \quad \parallel & \xrightarrow{\text{OH}^-} & | \\
\text{HC.CH}_2\text{.O.P.O.pantetheine} & & \text{4'-phosphopantetheine} + \text{C}=\text{CH}_2 \\
| \quad | & & | \\
\text{C}=\text{O} \quad \text{O}^- & & \text{C}=\text{O} \\
| & & | \\
\text{Leu} & & \text{Leu} \\
\qquad \text{(II)} & &
\end{array} \tag{1}
$$

$$
\begin{array}{c}
\text{Asp} \\
| \\
\text{N}-\text{H} \\
| \\
\text{C}=\text{CH}_2 \xrightarrow{\text{H}^+} \text{Asp} + \text{Leu} + \text{CH}_3\text{.CO.CO}_2\text{H} + \text{NH}_4^+ \\
| \\
\text{C}=\text{O} \\
| \\
\text{Leu}
\end{array} \tag{2}
$$

The structure of the peptide around the binding site is therefore:

O-4'-phosphopantetheine
|
Gly-Ala-Asp-Ser-Leu

The N-terminal residue of ACP is serine and the C-terminal residue is alanine.

The complete amino acid sequence of ACP from *E. coli* has now been established. It contains 77 residues in a single polypeptide chain with a high content of acidic residues (14 glu and 8 asp) but low in basic amino acids. This is reflected in its isoelectric point at pH 4.2. It has been noted that the sequence around the serine residue may be similar for the preparations that were tested from bacteria (*E. coli, Arthrobacter viscosus*) and also plant tissues (avocado and spinach).[10] Indeed, ACP derived from *E. coli* has been used successfully as a substrate for plant enzymes.

Studies on the reactivity of ACP from *E. coli* after tryptic hydrolysis indicate that at least two active sites are present: a substrate-binding site for attachment of the acyl group to 4'-phosphopantetheine and another region involved with binding ACP to β-oxoacyl-ACP reductase.[11] More probably, a number of further sites may be concerned in binding the polypeptide portion of the acyl-ACP derivatives to the individual enzymes. In the yeast and animal synthetases, an ACP-like peptide interacts structurally with the component enzymes to form unifunctional multienzyme complexes.

ACP and coenzyme A function as acyl group carriers in fatty acid synthesis and oxidation respectively; they both possess 4'-phosphopantetheine as prosthetic group. Vagelos and coworkers[12,13] have obtained evidence from pantothenate auxotrophs of E. coli that coenzyme A, known to be formed from 4'-phosphopantetheine, is the immediate precursor of ACP (equation (3)):

$$\text{4'-Phosphopantetheine} \xrightarrow{\text{2 ATP}} \text{dephospho-coenzyme A} \xrightarrow{\text{ATP}} \text{coenzyme A} \rightarrow \text{ACP} \quad (3)$$

These bacteria can utilize coenzyme A for ACP synthesis when transferred to a medium deficient in pantothenate. An enzyme, holo-ACP synthetase, that converts ACP-apoprotein and coenzyme A into ACP by transfer of the 4'-phosphopantetheine moiety has been isolated (equation (4)):

$$\text{Apo-ACP} + \text{coenzyme A} \xrightarrow{\text{Mg}^{2+}} \text{ACP} + \text{3',5'-adenosine diphosphate} \quad (4)$$

Protein-bound 4'-phosphopantetheine is also involved in reactions concerned in the synthesis of cyclic peptides by certain bacteria. Their mode of biosynthesis is particularly interesting and Lipmann[14] has drawn an analogy between this and fatty acid formation. Their synthesis proceeds by initial activation of the constituent amino acids into their corresponding acyl adenylates (as in ribosomal protein synthesis), followed by their conversion into the acyl thioester derivative and subsequent transpeptidation.

OVERALL SEQUENCE OF REACTIONS

The reactions involved in fatty acid synthesis in E. coli are presented in Scheme 3.1 (equation (5)–(11)). This sequence is very similar to that given by Lynen for the yeast synthetase, except that each reaction is catalysed by an individual enzyme that is not tightly bound to the remaining enzymes and

$$CH_3.CO.S.CoA + ACP.SH \rightleftharpoons CH_3.CO.S.ACP + CoASH \quad (5)$$

$$HO_2C.CH_2.CO.S.CoA + ACP.SH \rightleftharpoons HO_2C.CH_2.CO.S.ACP + CoASH \quad (6)$$

$$CH_3.CO.S.ACP + HO_2C.CH_2.CO.S.ACP \rightleftharpoons CH_3.CO.CH_2.CO.S.ACP + ACP.SH + CO_2 \quad (7)$$

$$CH_3.CO.CH_2.CO.S.ACP + NADPH + H^+ \rightleftharpoons$$
$$\text{D-}(-)\text{-}CH_3.CH(OH).CH_2.CO.S.ACP + NADP^+ \quad (8)$$

$$\text{D-}(-)\text{-}CH_3.CH(OH).CH_2.CO.S.ACP \rightleftharpoons CH_3.CH\overset{!}{:}CH.CO.S.ACP + H_2O \quad (9)$$

$$CH_3.CH\overset{!}{:}CH.CO.S.ACP + NADPH + H^+ \rightleftharpoons CH_3.CH_2.CH_2.CO.S.ACP + NADP^+ \quad (10)$$

$$CH_3.[CH_2]_{14}.CO.S.ACP + H_2O \rightarrow CH_3.[CH_2]_{14}.CO_2H + ACP.SH \quad (11)$$

Scheme 3.1   Mechanism of saturated fatty acid biosynthesis in bacteria may be purified. Similarly, the ACP derivatives remain free in solution but contain the substrates bound covalently by a thioester linkage. The acyl moieties of acetyl-CoA and malonyl-CoA react with ACP by means of transferase enzymes (equations (5) and (6)). Acetyl-ACP and malonyl-ACP thus formed are converted into acetoacetyl-ACP, with the release of CO_2, by the condensing enzyme (equation (7)). This product is then reduced with NADPH to give the D-$(-)$-β-hydroxybutyryl derivative which, in turn, is dehydrated to the α,β-unsaturated acyl-ACP. This is further reduced to the saturated butyryl-ACP (equations (8)–(10)). Butyryl-ACP undergoes a similar sequence of reactions which is repeated a further six times until the chain length reaches C_{16}. The terminal reaction (equation (11)) involves the hydrolysis of palmitoyl-ACP to the free acid. The other major acid in E. coli, cis-vaccenic acid, comprises approx. 70–80 per cent of the non-polar fatty acids and is also synthesized by this aggregate of enzymes. Its formation is identical with that of palmitic and stearic acid except that a cis double-bond is introduced at the β-hydroxydecanoyl-ACP stage by a specific β-hydroxy-decanoyl thioester dehydratase.

The isolation and characterization of the individual enzymes have been performed by Vagelos and Wakil and their respective coworkers in an elegant and exhaustive series of experiments.[15] The contribution from Bloch's laboratory has been directed mainly towards the study of the processes of unsaturation. All the enzymes of the E. coli synthetase have now been isolated and examined. Their properties are described in the following sections.

Acyltransferases

Both acetyl-CoA and malonyl-CoA transferase have been purified extensively from extracts of sonicated cells of E. coli.[6] Acetyl-CoA:ACP transferase is active with acetyl-CoA and other short-chain acyl-CoA compounds. Activity decreases as the chain-length increases up to octanoyl-CoA (approximately 10 per cent) and this parallels the fall in incorporation of these acyl-CoA compounds into fatty acids and hence their ability to act as primer for the overall process. Malonyl-CoA transferase is specific for this substrate. The purified enzyme was not inhibited by thiol group reagents but was sensitive to phenylmethanesulphonylfluoride, indicating the presence of an active serine residue.[16] In confirmation, malonyl enzyme was resistant to treatment with performic acid. Thus it was proposed, by analogy with previous data on the yeast and pigeon liver synthetases, that malonyl-CoA was transferred to ACP via an oxygen ester intermediate bound to the transferase (equations (12) and (13):

$$\text{Malonyl.S.CoA} + \text{enzyme.OH} \rightleftharpoons \text{malonyl.O.enzyme} + \text{CoASH} \qquad (12)$$

$$\text{Malonyl.O.enzyme} + \text{ACP.SH} \rightleftharpoons \text{malonyl.S.ACP} + \text{enzyme.OH} \qquad (13)$$

Condensing Enzyme (β-Oxoacyl-ACP Synthetase)

Soluble extracts were isolated from *C. kluyveri* and *E. coli* that catalysed the overall synthesis of fatty acids. Both systems catalysed the condensation reaction,[5,1] as determined by the malonyl-CoA–CO_2 exchange reaction, after measuring the amount of $H^{14}CO_3^-$ fixed into acid-stable material. Purified preparations were obtained from sonicated cells of *E. coli* that were capable of catalysing the condensation of malonyl-ACP with acetyl-ACP or the short-chain homologues.[17] This reaction is therefore responsible for chain elongation. It was assayed by coupling to the β-oxoacyl-ACP reductase and the overall rate was determined by measuring the decrease in extinction at 340 nm in the presence of NADPH. Other condensing enzymes may be present that are specific for the longer chain lengths (C_{10}–C_{16}). The β-oxoacyl-ACP thioesters were further identified by their characteristic extinction at 303 nm and by their behaviour on Sephadex and DEAE-cellulose. The enzyme was inactive with acetyl-CoA.

The condensing enzyme possesses a functional thiol group that is susceptible to the action of thiol reagents. Acetyl-ACP and other short-chain derivatives protect the enzyme from these inhibitors, but malonyl-ACP, malonyl-CoA and ACP are inactive in this respect. Thus acyl-ACP derivatives appear to interact with the enzyme, possibly forming an acyl. S. enzyme prior to reaction. This situation is evidently analogous to the transfer of acyl thioesters to the 'peripheral' site within the condensing enzyme component of the yeast and animal synthetases before reaction with the malonyl group at the 'central' pantetheine site (see Chapter 2; Scheme 2.5).

This enzyme is unusual in that it catalyses the interaction of two substrates, both of which are bound to a carrier protein molecule with molecular weight of 9,000.[17] The enzyme itself has a molecular weight of 66,000. The reaction probably occurs in two stages[17,18] (equations (14)–(15)):

$$R.CO.S.ACP + HS.enzyme \rightleftharpoons R.CO.S.enzyme + ACP.SH \qquad (14)$$

$$R.CO.S.enzyme + HO_2C.CH_2.CO.S.ACP \rightleftharpoons R.CO.CH_2.CO.S.ACP + CO_2 + enzyme.SH \qquad (15)$$

A cysteine residue is implicated as the thiol binding site since alkylation by iodoacetamide results in inhibition that is prevented by addition of acetyl-ACP.[18] Moreover, the initial product of the reaction has been identified as acetyl.S.enzyme whose properties (destruction by alkali, hydroxylamine or performic acid) indicate that the acetyl group is bound to a cysteine residue. Some physico-chemical characteristics of the highly purified enzyme have been described recently.[19]

β-Oxoacyl-ACP Reductase

This enzyme exhibits a wide specificity and acts on the appropriate ACP derivatives with chain-length ranging from C_4–C_{16}, in the presence of

NADPH, to give the D-(−)-stereoisomer of the β-hydroxy acyl-ACP thioester.[14] NADH is inactive in catalysing this reduction. The equilibrium for the reaction greatly favours the synthesis of the β-hydroxyacyl-ACP and hence fatty acid synthesis.[20] Highly purified preparations of β-oxoacyl-ACP reductase are unstable but inactivation may be prevented by addition of NADPH, or multivalent anions. The acyl substrates are bound to the enzyme and react as the β-oxo form.[21]

Dehydratase

Several enzymes with dehydratase activity have been prepared from *E. coli*.[22] A non-specific enzyme with regard to chain-length has been purified and acts upon D-(−)-β-hydroxyacyl-ACP compounds giving rise to the α,β-unsaturated acyl-ACP products.[23] It is inactive with the corresponding L-(+)-stereoisomers or coenzyme-A derivatives. Thus, D-(−)-β-hydroxy-butyryl-ACP is the only isomer involved in fatty acid synthesis in the systems studied to date, regardless of source. A short-chain dehydratase is active for the C_4 to C_8-ACP derivatives with optimum activity on β-hydroxy-butyryl-ACP.

β-Hydroxydecanoyl-ACP Dehydratase

Wakil, Bloch and Vagelos and their groups have also examined a certain dehydratase that forms *trans*-2-decenoyl-ACP and is specifically implicated in the synthesis of saturated fatty acids. This enzyme is also active for the C_{12}–C_{16} substrates.[24,25]

Bloch and coworkers, however, have isolated a similar enzyme from *E. coli* that catalyses the dehydration of β-hydroxydecanoyl-ACP but, in addition, is active with the coenzyme A, pantetheine and *N*-acetylcysteamine derivatives.[26] This study proved particularly rewarding since the significant product in this case is *cis*-β,γ-decenoyl-ACP, which is formed together with the *trans*-α,β-derivative. The *cis* double bond is not susceptible to reduction and is retained throughout subsequent chain elongation (after condensation with malonyl-ACP, reduction, etc.) to give ultimately *cis*-vaccenate ((III); see Scheme 3.2). The enzyme is responsible for the synthesis of unsaturated fatty acids by anaerobic bacteria but it also makes an important contribution under aerobic conditions.

The characteristics of this reaction have been examined using β-hydroxy-decanoyl-*N*-acetylcysteamine as model substrate together with the highly purified enzyme.[27] The products were identified as the *trans*-2-decenoyl and *cis*-3-decenoyl-derivatives.[28] The enzyme was multifunctional and catalysed the interconversions of β-hydroxydecanoyl-*N*-acetylcysteamine (70 per cent), *trans*-2-decenoyl (27 per cent) and *cis*-3-decenoyl (3 per cent) derivatives; the approximate equilibrium concentrations are given in parenthesis.[29] The dehydratase and isomerase functions of this enzyme remained constant during extensive purification. Experiments with labelled substrates and the evaluation of $^3H/^{14}C$ ratios were used to establish the reaction mechanism of this dehydratase[30] (equation (16)):

$$\beta\text{-Hydroxydecanoyl.S.X}$$
$$cis\text{-}\beta,\gamma \;\rightleftharpoons\; (trans\text{-}\alpha,\beta)\text{enzyme} \;\rightleftharpoons\; trans\text{-}\alpha,\beta \qquad (16)$$

The direct formation of the β,γ-decenoyl derivative by dehydration would not be kinetically favourable since the C–H bond in the C_γ position possesses negligible acidity compared with the equivalent bond at C_α. Thus, the cis-3-decenoyl (β,γ) derivative is formed by isomerization of the enzyme-bound trans-2-decenoyl compound.[28]

The relative predominance of cis-vaccenate as the final product in the synthetase reaction and in vivo, despite the apparent favour in the direction of the α,β-derivative at this dehydratase level, has been explained[30] by suggesting that this enzyme is not entirely responsible for the eventual proportion of cis to trans, that is, unsaturated to saturated acids.[31] This may be related to competition between the rate of condensation of β,γ-decenoyl-ACP with malonyl-ACP on the one hand and reduction of α,β-decenoyl-ACP on the other (Scheme 3.2). If the rate of the former reaction is more rapid, it will compensate for the low equilibrium concentration and tend to favour production of the C_{12} unsaturated ACP and ultimately cis-vaccenate.

The site of differentiation leading to unsaturation at the level of this dehydratase has been confirmed by studying a mutant of E. coli that requires unsaturated fatty acids for growth.[25] The enzyme defect responsible was identified as the lack of β-hydroxydecanoyl-ACP dehydratase.[32] However, this organism did synthesize saturated fatty acids normally. This situation implicated the existence of a further dehydratase that is responsible for the formation of α,β-decenoyl-ACP (and higher homologues) but which does not possess an isomerase function. It is clear therefore that the presence of the specific C_{10} dehydratase is essential for the formation of cis-vaccenate and the related palmitoleate in E. coli and other bacteria. A scheme of reactions incorporating Bloch's results is presented (Scheme 3.2).

Enoyl-ACP Reductase

This enzyme catalyses the reduction of α,β-unsaturated acyl-ACP thioesters to their saturated derivatives.[33] It has been extensively purified and two enzymic activities have been resolved, one requiring NADPH as hydrogen donor and the other NADH. No flavin component has been detected. The NADPH-requiring enzyme is optimally active at the C_4 and C_6 level and is absolutely specific for ACP substrates. It is inactive above pH 8·0. The NADH reductase catalyses reactions with thioesters of ACP or coenzyme A with maximum activity for the C_6 and C_8 derivatives but

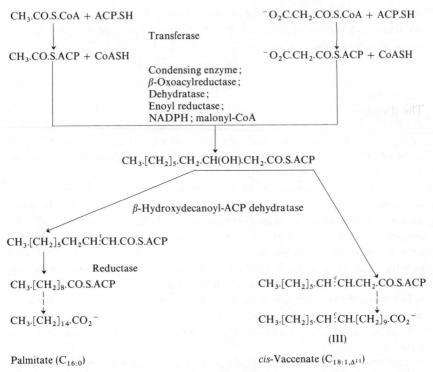

Scheme 3.2   Representation of the reactions leading to the synthesis of palmitate and
cis-vaccenate

considerable activity is also exhibited by this enzyme for the longer chain
substrates. It has a broad pH optimum (6·0–9·0) unlike its NADPH-requiring
counterpart. Both preparations possess functional thiol groups. Although
two enzyme proteins have not yet been resolved it is thought that the reductase activities reside in two distinct enzymes. An interesting parallel may be
noted between this requirement for NADH and the similar property in the
enoyl reductase step involved in microbial synthetases[34] and in the mitochondrial elongation of fatty acids within animal tissues (Chapter 2).

Control Aspects

Very little work has been conducted on these aspects of bacterial fatty
acid synthesis compared with the vast range of investigations in animal
systems (Chapter 4). However, addition of monounsaturated acids to growing
cells of L. plantarum was found to inhibit the synthesis of the major fatty
acids, principally palmitate, cis-vaccenate and lactobacillate.[35] Oleate and
cis-vaccenate proved the most effective in this respect and caused a severe
reduction in the activity of acetyl-CoA carboxylase and the individual

enzymes of the fatty acid synthetase (but not palmitoyl thioesterase). This inhibition was due to repression of enzyme synthesis rather than a more direct effect on enzyme activity.[35]

The exogenous unsaturated acids were readily incorporated into lipids with consequent increases in the unsaturated acid content of phospholipids in membranes.

Palmitoyl Thioesterase

This enzyme (palmitoyl-ACP hydrolase) may be responsible for the hydrolysis of the terminal long-chain acyl-ACP, with the release of free fatty acid and ACP. Two distinct enzymes have been isolated and purified from *E. coli*; one acts optimally on the acyl thioesters (ACP and coenzyme A derivatives) of palmitic, palmitoleic and *cis*-vaccenic acids but will also hydrolyse shorter-chain length derivatives.[36] Preparations of synthetase obtained from certain bacteria that are capable of forming stearate (instead of the more usual palmitate) synthesize palmitate on supplementation with this enzyme. A second enzyme with more general activity (C_6–C_{18}, including β-hydroxy-derivatives) also utilizes C_{16} as the preferred chain-length.[37]

However, it has been shown that palmitoyl-ACP may be used directly for lipid synthesis (Chapter 5) without prior hydrolysis by transfer to glycerol 3-phosphate. The presence of a thioesterase may therefore be incidental to that of fatty acid synthetase. Moreover, this enzyme is not repressed coordinately with other enzymes of the synthetase group under conditions in which their synthesis is markedly reduced.[35]

Reconstitution of the Bacterial Fatty Acid Synthetase

Wakil and coworkers[23] have investigated the properties of the *E. coli* synthetase that had been reconstituted from the individual purified enzymes. Efficient incorporation of [^{14}C]malonyl-CoA into fatty acids depended on the presence of acetyl-ACP, NADPH, NADH and the five enzymes concerned, especially the condensing enzyme, β-oxoacyl-ACP reductase and the dehydratase (Table 3.1). The products were converted into their methyl

Table 3.1 Enzymic requirements for the reconstituted fatty acid synthetase (Mizugaki *et al.*, 1968)[23]

Enzyme omitted	[^{14}C]Malonyl-CoA incorporated (counts/min)
None (complete system)	19,900
Malonyl-CoA:ACP transferase	9,700
Acyl-malonyl-ACP condensing enzyme	300
β-Oxoacyl-ACP reductase	600
β-Hydroxyacyl-ACP dehydratase	1,200
Enoyl-ACP reductase	6,500

esters and resolved by thin-layer chromatography, using silver nitrate impregnated plates into hydroxy (84 per cent), saturated (9 per cent) and unsaturated (7 per cent) acids. The major component of the hydroxy acid fraction was identified as β-hydroxydecanoate.

Moreover, two fractions (termed A and B) could be isolated from the synthetase preparation which possessed long-chain dehydratase activity but had low activity with regard to the overall process of fatty acid synthesis. Accumulation of β-hydroxydecanoyl-ACP in the reconstituted system is due to the apparent lack of activity by these dehydratases and inability of the short-chain dehydratase to act on this C_{10} substrate. Addition of enzyme (A) to the system stimulated net fatty acid synthesis and also caused a large increase in saturated acids at the expense of the hydroxy products. The other enzyme (B) also stimulated total synthesis and, at the same time, increased the conversion of the β-hydroxy compound into both saturated and unsaturated acids (Table 3.2). Thus, enzyme B has the ability to dehydrate

Table 3.2   Effect of addition of enzyme fractions to a reconstituted fatty acid synthetase (Mizugaki et al., 1968)[23]

Enzyme preparation	Total radioactivity (counts/min)	[^{14}C]Malonyl-CoA incorporated (Percent) radioactivity in		
		Hydroxy	Saturated	Unsaturated
None	100	—	—	—
Reconstituted system	4,000	84	9	7
Enzyme (A)	1,800	50	42	8
Enzyme (B)	140	—	—	—
Reconstituted system + (A)	9,800	48	44	8
Reconstituted system + (B)	7,200	19	32	49

β-hydroxydecanoyl-ACP to both cis-β,γ- and trans-α,β-decenoyl-ACP intermediates and is presumably similar to that previously described by Bloch.[27] Enzyme (A) forms mainly the trans-product and hence favours the formation of saturated acids.

FATTY ACID SYNTHESIS IN PLANTS

The commonly occurring fatty acids in plants are the $C_{16:0}$, $C_{18:1}$ and $C_{18:2}$ acids but saturated acids of greater chain-length (C_{18}–C_{24}) are localized in the wax fraction of surface lipids. The details of the reactions involved for de novo synthesis are essentially similar to those found for other systems and they have been examined mainly by Stumpf and colleagues. Plant synthetase preparations have been resolved into a number of active fractions that may subsequently be reconstituted. A common component in plant and bacterial systems is a freely dissociable acyl carrier protein.

Stumpf and Barber[38] first isolated a cell-free particulate preparation from avocado mesocarp that catalysed the synthesis of palmitate, stearate and oleate from acetate, and their esterification with glycerol, in the presence of ATP, coenzyme A, CO_2 and Mn^{2+}. Degradative studies established that these acids had been formed by means of *de novo* synthesis. The particles were designated as mitochondria since they possessed activities characteristic of β-oxidation, tricarboxylic acid cycle reactions and oxidative phosphorylation. A water-soluble system derived from mitochondria was subsequently isolated that was capable of converting acetyl-CoA plus malonyl-CoA into palmitate and stearate.[39] No evidence could be obtained in this work for a soluble enzyme complex in the cytoplasm, in apparent contrast with other systems tested (animal, yeast and bacteria) but this was later attributed to an absolute requirement for exogenous ACP.[40]

The fatty acid synthetase from this particulate fraction of avocado mesocarp has since been resolved into two components.[41] A heat-stable protein and heat-labile fraction were involved and both were essential for activity. The heat-stable protein could be replaced with ACP from *E. coli*. Lettuce and spinach chloroplasts also yielded active extracts that converted malonyl-CoA into stearate in the presence of NADPH, and ACP from *E. coli* or plant sources.[42] The enzymes appeared to be related to the stromal proteins and did not involve the grana; they remained in the high-speed supernatant after centrifuging crushed chloroplasts at 105,000 g for 25 min.[42,43] Even after 5 h centrifugation, only 10 per cent of the activity sedimented out. Intact chloroplasts, however, synthesized large amounts of unsaturated acids with oleate predominating among the $C_{18:1}$ acids.[44] Attempts were made to resolve the lettuce synthetase into its component enzymes by means of isoelectric precipitation but only limited success was achieved.[43] Two fractions were obtained; the supernatant was completely inactive but the precipitated protein did retain some activity which was much increased on recombination with the supernatant material. However, the total recovery was poor since only a small amount of the original activity was regained.

There have been indications that fatty acid synthesis from acetyl-CoA in spinach chloroplasts may occur in the particulate fraction.[45] Possibly this represents a different system from that catalysed by the soluble proteins that utilizes malonyl-CoA; this may prove to be similar to the chain elongation mechanism found in animal mitochondria.

A procedure for the preparation of mitochondrial and supernatant fractions from avocado mesocarp, formed after centrifugation on a sucrose density gradient, has been briefly described.[46] These fractions were apparently inactive in effecting fatty acid synthesis but the more dense chloroplasts incorporated acetate and malonate into palmitate and oleate. The implications of these findings were that previous incorporations achieved with mitochondrial and supernatant extracts resulted from mechanical disruption of the chloroplasts with subsequent leakage and contamination. However, the fact that exogenous ACP was not added to these systems might explain the negative results.[40]

The aforementioned work all relates to tissues of high lipid content that may be expected to contain synthetases of high specific activity. An active synthetase preparation has also been isolated, however, from soluble extracts of potato tuber, a starch-storage organ with low lipid content.[47] Products ranged from the saturated C_8-C_{16} acids with the long-chain acids predominant under conditions of excess malonyl-CoA and NADPH. These extracts and particle-free supernatants derived from a number of plant sources required the addition of ACP for effective fatty acid synthesis.[40]

FATTY ACID SYNTHETASES FROM *MYCOBACTERIUM PHLEI* AND *EUGLENA GRACILIS*

Bloch and colleagues[48,49] have recently presented their interesting results concerning the characterization of fatty acid synthetase complexes from *M. phlei*, one of the more advanced prokaryotic organisms, and etiolated (dark-grown) cells of the phytoflagellate *E. gracilis*.[50,51] Prior to this work, two groups of synthetase had been described: the multienzyme complexes of high molecular weight found in yeasts and animal tissues, and the multienzyme aggregates derived from bacteria and plants. The *M. phlei* complex, however, had a molecular weight 1.7×10^6 and behaved as a unifunctional entity that did not need exogenous ACP for activity.[48] The principal products were the C_{18} and C_{24} acids, reflecting the pattern of the fatty acids in the lipids of intact cells. Acetyl-CoA, octanoyl-CoA and stearoyl-CoA were all capable of priming the system effectively. The nature of the fatty acids formed was qualitatively similar except that stearoyl-CoA gave acids with chain-length C_{20} and above. Enzymic activity was dependent upon the integrity of this bacterial complex and was rapidly lost in buffers of low ionic strength. Another feature of the complex was that an external heat-stable supplement was required. This comprised of two components, FMN (for the second reductase reaction) and polysaccharide which exerts its activity by greatly reducing the K_m of the synthetase for acetyl-CoA.[49] In this respect its behaviour differs from that of other synthetase units which are self-sufficient in terms of bound cofactor requirements.

Data concerning the synthetase complex from etiolated cells of *E. gracilis* are also available.[50] This cytoplasmic multienzyme unit had a high molecular weight and yielded mainly palmitate as product. Photoauxotrophic cells (grown in the light) provided, in addition, a synthesizing system associated with the chloroplasts that was made up of individual enzymes that did not aggregate and required a supply ACP for activity. It gave stearate and arachidate (C_{20}) as principal products. Illumination of etiolated cells induced the formation of the enzymes responsible for the latter synthetase but this could be suppressed by addition of chloramphenicol (but not cycloheximide) as inhibitor of protein synthesis.[51] Derivation of the ACP-dependent system occurred in the absence of net growth and required *de novo* formation of enzymes. The provision of different types of synthetase illustrates the capacity of *E. gracilis* to adapt to environmental changes, an ability that is also expressed in the behaviour of the enzymes responsible for the synthesis of unsaturated acids. This will be discussed later (p. 56).

COMPARATIVE COMPOSITION AND ACTIVITIES OF ACYL CARRIER PROTEINS AND SYNTHETASES

ACP's from *E. coli*, *A. viscosus* and two higher plants (avocado mesocarp and spinach chloroplasts) have been purified and analysed for their constituent amino acids.[10] In general, the composition of these four proteins is quite similar; they have approximately the same molecular weight (9,000–12,000) and possess high levels of the acidic aspartic and glutamic acids. They are all heat- and acid-stable. The amino acids at the central region that binds 4'-phosphopantetheine were isolated from these products after tryptic digestion and seem to be identical.[52] The prosthetic group is covalently attached by means of a phosphodiester link to a serine residue. The ACP from *C. butyricum* has not been examined as rigorously as those mentioned above but has molecular weight 8,100 and contains one mole of 4'-phosphopantetheine.[53] Lynen's group[54] have compared similar data for yeast ACP (Chapter 2, p. 23).

Effect of Replacement of Bacterial/Plant ACP's

The bacterial ACP's are functionally similar to each other as are the plant counterparts. Moreover, addition of ACP obtained from *E. coli* to the spinach synthetase was effective for fatty acid synthesis and the type of acid produced was not changed. However, substitution of the plant ACP in the *E. coli* system stimulated the production of β-hydroxy derivatives of intermediate chain-length (saturated and monoenoic). This phenomenon was explained by suggesting that the plant β-hydroxy acyl-ACP intermediates were defective substrates for the α,β-dehydratase that acts beyond the C_{10} stage.[10] The specific C_{10}-β,γ-dehydratase also appeared to be far less active with substrates bound to the plant ACP and there was a consequent displacement towards the synthesis of saturated acids. The plant synthetase did not contain this particular β-hydroxydecanoyl-ACP dehydratase.[52] Another possibility that explains the modification in the nature of the products, however, may lie in the relative binding properties of the peripherally located amino acids of the plant and bacterial ACP, as against the apparently similar central binding region.[52]

Further progress in the relation between structure and function of ACP may now become more rapid since its apo form, identical with the *E. coli* product in the functional residues, has been synthesized in fair yield by the solid phase method.[55] This material was converted into ACP by enzymic addition of the prosthetic group (equation (4)). The product cochromatographed with authentic material and possessed malonyl pantetheine-CO_2 exchange activity.

Aggregation of Enzymes in Multienzyme Complexes

Fatty acid synthetases invariably catalyse the conversion of acetyl-CoA and malonyl-CoA into long-chain fatty acids. Bloch and coworkers,[48] in describing the isolation of synthetases from *M. phlei* and *E. gracilis*, discuss the characteristics of the two major groups of fatty acid synthesizing systems. Lynen *et al.*[56] previously commented on the probable advantage to biological systems that is gained from an aggregation of enzymes behaving as a single operational unit. Possibly, this represents a more advanced stage in the evolution of the enzymic sequence. Presumably the organized structure is catalytically more efficient, since the transfer of the intermediates among the enzymes could occur without a need for free diffusion or dilution with the cytoplasmic contents. It has been calculated[56] that one mole of yeast synthetase may convert 6,000 moles of malonyl-CoA into fatty acids per min and the local concentration of intermediates at any instant may reach a value of 5×10^{-4} M. However, it should be realized that the bacterial (and plant) synthetases are also exceedingly efficient (acyl-ACP intermediates do not accumulate during the synthesis) and could be organized into a less tightly bound complex *in vivo*. These may be more readily dissociated into their component enzymes during the isolation procedures or subsequent fractionation.[57] Indeed, data gained after electron microscopy and autoradiography of suitable preparations of *E. coli* indicated that ^{3}H-labelled ACP (derived from β-^{3}H-alanine) was localized at a site close to the cytoplasmic membrane in this organism[58] suggesting that the bacterial synthetase may possess some degree of organization *in vivo*. Moreover, this situation would exercise great efficiency in the synthesis of membrane phospholipids from ACP-linked precursors that were formed in this region of the cell, since their availability would be readily enhanced.

In both groups of synthetase the intermediates are bound to ACP, a protein analogue of coenzyme A and are therefore rendered inaccessible to competing reactions.[54] The use of this cofactor for synthesis versus the coenzyme A derivative in the oxidative process lies in the fact that it enables the cell to distinguish between intermediates (that are essentially the same) in fatty acid metabolism.[9,56]

MECHANISMS FOR THE SYNTHESIS OF UNSATURATED ACIDS IN BACTERIA AND PLANTS

In anaerobic bacteria, the process of unsaturation is clearly independent of oxygen and a mechanism has evolved for this purpose that is closely related to *de novo* synthesis. A single double bond is inserted at the C_{10} level by dehydration of β-hydroxydecanoyl-ACP to give *cis*-3-decenoyl-ACP; this unsaturation is retained in the final product, *cis*-vaccenate $(C_{18:1,\Delta^{11}})$. A similar reaction at the *cis*-3-dodecenoyl-ACP stage would give rise to oleate (Δ^9). This acid occurs to a far lesser extent in bacteria and is generally formed by an aerobic mechanism. It is completely absent from *E. coli*.[59]

Aerobic Unsaturation in Bacteria

Bloch and colleagues[60,61] investigated the production of oleate in *M. phlei* and showed that it was derived from preformed stearate under aerobic conditions. Experiments with cell-free particles confirmed that palmitoyl-CoA and stearoyl-CoA were both converted into the *cis*-9,10-unsaturated derivatives.[62] The process required O_2, NADPH and, in addition, Fe^{2+} and catalytic amounts of flavin.

Similar results have been obtained with a limited number of species from different genera of bacteria.[63] *Bacillus megaterium*, however, is even more atypical in that it produces *cis*-Δ^5-C_{16} and -C_{18} acids. These particular bacteria are all unusual in synthesizing monoenoic fatty acids *via* an oxygen-dependent process since the alternative method linked to *de novo* synthesis is so readily available. With regard to the mechanism entailed, *Corynebacterium diphtheriae* converts stearate into oleate by means of stereo-specific removal of the D-hydrogen atoms attached to C-9 and C-10. The four monotritiated isomers of stearate labelled at C-9 and C-10 were used as test substrates.[64]

The first report concerning the presence of dienoic acids in bacteria has only recently appeared.[65] Two hexadecadienoic acids were formed from palmitate by *B. licheniformis* when incubated at a temperature that was rather lower than the usual value. The enzyme system responsible for the insertion of the double bond at C-5 in the synthesis of the Δ^5-monoenoic and $\Delta^{5,10}$-dienoic acids was temperature-sensitive and only present at the lower temperature.

Unsaturation in Plants

Mudd and Stumpf[66] were the first to show that the formation of oleate in a plant tissue was completely dependent on oxygen. In its absence, conversion of radioactivity from [^{14}C]acetate into stearate by avocado mesocarp was correspondingly increased at the expense of oleate formation. This work was followed by the demonstration that *E. gracilis* might employ two systems under different physiological circumstances.[67] This organism possesses an oxygen- and NADPH-dependent dehydrogenating mechanism when grown in the dark (etiolated form) that is specific for acyl-CoA thioesters. However, when it is grown in the presence of light (photoauxotrophic), the preferred substrate for the purified desaturase may be either stearoyl-CoA or stearoyl-ACP, the product formed during fatty acid synthesis.[68] In both cases, unsaturation occurs at the same position in the stearoyl residue with the production of the *cis*-9-isomer. The explanation for this alteration in requirement of the thioester portion of the substrate is not understood but it may simply reflect variations in permeability factors or the presence of a suitable acyl transferase under the different growth conditions. A soluble preparation derived from chloroplasts of photoauxotrophic cells was subsequently resolved into three essential components: a flavin-containing NADPH-oxidase, non-haem iron (ferredoxin) and a desaturase.[69]

A similar ferredoxin-dependent system has been isolated from spinach chloroplasts which also uses stearoyl-ACP. It is interesting to note here that the ACP used in these experiments was isolated from *E. coli* since it is easier to obtain from this source. Further investigations established that 9- and 10-hydroxystearoyl-ACP were inactive as substrates,[68] militating against the possible involvement of a monooxygenase enzyme and confirming the role of the desaturase enzyme.

Similar results showing a dependency on oxygen and a reduced nucleotide cofactor were obtained with photosynthetic cultures of the alga *Chlorella vulgaris*. Linoleate ($C_{18:2,\Delta^{9,12}}$) and α-linolenate ($C_{18:3,\Delta^{9,12,15}}$) were formed aerobically from oleoyl-CoA by chloroplast preparations.[70] Conversion of stearate into oleate occurred by stereospecific removal of the D-9 and D-10 atoms of the deuterated precursor (that is *cis*-elimination, c.f. *C. diphtheriae*); further removal of hydrogen atoms attached to C-12 and C-13 (also in the D-configuration) and C-15 and C-16 resulted in the formation of linoleate and α-linolenate respectively.[71] These results may be seen in Scheme 3.3.

$$CH_3.[CH_2]_{16}.CO.S.CoA \xrightarrow[\Delta^9\text{-desaturation}]{\text{loss of H-9, H-10}} {}^{18}CH_3.[CH_2]_7.CH:{}^9CH.[CH_2]_7.{}^1CO.S.CoA$$
$$C_{18:0} \qquad\qquad C_{18:1,\Delta^9}$$

$$\xrightarrow[\Delta^{12}\text{-desaturation}]{\text{loss of H-12, H-13}} {}^{18}CH_3.[CH_2]_4.CH:{}^{12}CH.CH_2.CH:{}^9CH.[CH_2]_7.{}^1CO.S.CoA$$
$$C_{18:2,\Delta^{9,12}}$$

$$\xrightarrow[\Delta^{15}\text{-desaturation}]{\text{loss of H-15, H-16}} {}^{18}CH_3.CH_2.CH:{}^{15}CH.CH_2.CH:{}^{12}CH.CH_2.CH:{}^9CH.[CH_2]_7.{}^1CO.S.CoA$$
$$C_{18:3,\Delta^{9,12,15}}$$

$$-CH_2.CH_2- + NADPH + H^+ + O_2 \rightarrow -CH:CH- + NADP^+ + 2H_2O$$

Scheme 3.3 Aerobic sequence leading to the introduction of sites of *cis* unsaturation in long-chain acyl thioesters

Further details concerning the second Δ^{12}-desaturation step have been revealed with the use of a purified enzyme obtained from microsomes of maturing safflower seeds.[72] This converted oleoyl-CoA directly into the dienoic linoleoyl-CoA thioester prior to transfer to an endogenous acyl acceptor lipid to give phosphatidyl choline. The β-oleoyl phospholipid derivative (in which the oleoyl residue was attached as an oxygen ester) did not behave as a substrate although this product and the linoleoyl derivative were rapidly formed under the incubation conditions through the agency of an acyl transferase. It is thought that this mechanism involving the CoA form of the $C_{18:1}$ acid is generally applicable to other non-plant systems.

CYCLOPROPANE AND C-METHYL FATTY ACIDS IN BACTERIA

Cyclopropane Acids

This type of acid of chain-length C_{17} and C_{19} occurs in relatively high concentration in the membrane lipids of numerous bacteria, especially the lactobacilli. They are rarely found elsewhere in Nature but are occasionally present in certain plant oils. The first such acid that was isolated was termed lactobacillic acid and was characterized as cis-11,12-methyleneoctadecanoic acid (IV).[73] The related C_{17} homologue, cis-9,10-methylenehexadecanoic acid, has been isolated from E. coli.[74] It was proposed that lactobacillate was formed by the addition of a C_1 unit across the double bond at C-11 of cis-vaccenate (III) (equation (17)):

$$CH_3.[CH_2]_5.CH:CH.[CH_2]_9.CO_2^- \xrightarrow{C_1} CH_3.[CH_2]_5.\overset{\displaystyle CH_2}{\overset{\diagup\diagdown}{CHCH}}.[CH_2]_9.CO_2^- \quad (17)$$

$$\text{(III)} \qquad\qquad\qquad\qquad\qquad\qquad \text{(IV)}$$

cis-Vaccenate was indeed incorporated intact in L. arabinosus[75] and it was later shown with extracts from various bacteria that the C_1 unit was derived from the methyl group of S-adenosylmethionine.[76,77] Endogenous lipids present in the incubation mixture were essential for incorporation of activity.[78] The radioactive product was identified as a phosphatidylethanolamine. It was later confirmed[79] that an olefinic linkage in phosphatidylethanolamine (with preferential specificity for the 1-acyl position) was the true acceptor of the additional C_1 unit (equation (18)) but this substrate had to be suitably dispersed for optimum activity:

$$
\begin{array}{l}
^1CH_2.O.CO.[CH_2]_7.CH:CH.[CH_2]_7.CH_3 \\
^2CH.O.CO.R \\
\quad\quad\; O \\
\quad\quad\; \| \\
^3CH_2.O.\overset{}{P}.O.CH_2.CH_2.NH_3^+ \\
\quad\quad\; O^-
\end{array}
\xrightarrow{\text{S-adenosylmethionine}}
\begin{array}{l}
\overset{\displaystyle CH_2}{\overset{\diagup\diagdown}{}} \\
CH_2.O.CO.[CH_2]_7.CH.CH.[CH_2]_7CH_3 \\
CH.O.CO.R \\
\quad\quad\; O \\
\quad\quad\; \| \\
CH_2.O.\overset{}{P}.O.CH_2.CH_2.NH_3^+ \\
\quad\quad\; O^-
\end{array}
\quad (18)
$$

The 'cyclopropane synthetase' exhibited an exclusive specificity for the 3-phosphoglycerol derivative, that is, the natural substrate.

The mode of synthesis of cis-9,10-methylenehexadecanoate has also been studied by Law and coworkers[80] with particular reference to the status of the hydrogen atoms originating from methionine. [methyl-$^2H_3$]Methionine was used as labelled substrate but only two deuterium atoms were incorporated from the methyl group into the C_{17} acid. Earlier work on the mechanism of biological methylation reactions had indicated that transfer of the C_1 unit from S-adenosylmethionine occurred after electrophilic attack by the $CH_3.S:^+$ group on a nucleophilic group such as hydroxyl, amino or $C:C$ within the substrate molecule.[80] Akhtar et al.[81] proposed (with respect to ergosterol biosynthesis) that the entire methyl group of S-adenosylmethionine was transferred intact to an olefinic group in the sterol side-chain, resulting in the formation of a carbonium ion. A 1,2-hydride shift could then occur to stabilize the product with loss of one of the hydrogen atoms of the methyl group to give a cyclopropane or methylene derivative.

C-Methyl Acids

These acids are also prevalent in mycobacteria and are characteristic of membrane lipids. Again, the additional C_1 unit is inserted across a double bond in a suitable substrate,[61] followed by reduction. The C-9 and C-10 atoms in oleate react in this way to give tuberculostearate (10-methyl-stearate, (V)) (equation (19)):

$$CH_3.[CH_2]_7.CH:CH.[CH_2]_7.CO_2^- \xrightarrow{C_1} CH_3.[CH_2]_7.\overset{\overset{\displaystyle CH_3}{|}}{C}H.CH_2.[CH_2]_7.CO_2^- \quad (19)$$

$$(V)$$

Synthetically prepared 10-methylenestearate is also effectively incorporated into this product and acts as an intermediate in the process. More recently the endogenous lipid acceptor has been identified as an olefinic residue within a phospholipid (as had previously been shown for cyclopropane acids) and the methyl donor as S-adenosylmethionine.[82] The reaction occurs in two stages: formation of a 10-methylene-derivative followed by an NADPH-mediated reduction to the C-methyl phospholipid (equations (20) and (21)):

$$-^{10}CH:^9CH- + \text{S-adenosylmethionine} \rightleftharpoons -\overset{\overset{\displaystyle CH_2}{||}}{C}.CH_2- + \text{S-adenosylhomocysteine} \quad (20)$$

$$-\overset{\overset{\displaystyle CH_2}{||}}{C}.CH_2- + NADPH + H^+ \rightleftharpoons -^{10}\overset{\overset{\displaystyle CH_3}{|}}{C}H.^9CH_2- + NADP^+ \quad (21)$$

The enzymes responsible were present in the particulate and supernatant fractions of sonicated cells of M. phlei, indicating that the natural site of the enzymes may be localized in the membrane. Various phospholipids were active as substrates but only the Δ^9-position within the unsaturated acid group was sensitive to alkylenation.

Lederer and his colleagues,[83,84] in the course of an elegant series of investigations with deuterated methionine, have established the mode of transfer of the methyl group to acceptor olefinic and also phenolic and enolic groups. Tuberculostearate (V) is formed in *M. smegmatis* from [*methyl*-$^2H_3$]-methionine with retention of two deuterium atoms from the original substrate. The labelled product was examined as its methyl ester by mass spectrometry; this procedure showed conclusively that only two deuterium atoms were retained. Examination of the peaks confirmed that the deuterium atoms were located in the methyl group attached to C-10 of the molecule.

Lederer[83] has suggested that branched-chain and methylene-bridged acids perform the same physiological function as the related unsaturated fatty acids. The lower melting point that results confers certain desirable features on the lipids and hence the membranous structures into which they are incorporated. However, this teleological explanation for their synthesis may not be correct, since both cyclopropane and *C*-methyl acids are formed from the related unsaturated derivatives. On the other hand, removal of the unsaturation permits different folding patterns to be achieved by the aliphatic acyl residues. Mycobacteria are also capable of synthesizing polymethylated acids such as the C_{32} mycocerosic acid (VI) from a normal C_{20} primer plus four propionyl residues (presumably derived from methylmalonyl-CoA).[85] In addition, methyl-branched acids of the iso and anteiso type are derived in part from the catabolic products of branched-chain amino acids.

$$CH_3.[CH_2]_{18}.CH_2.[CH(CH_3).CH_2]_3.CH(CH_3).CO_2H$$

(VI)

REFERENCES

1. Goldman, P., Alberts, A. W., and Vagelos, P. R., *J. Biol. Chem.*, **238**, 1255, 3579 (1963)
2. Lennarz, W., Light, R., and Bloch, K., *Proc. Nat. Acad. Sci. U.S.*, **48**, 840 (1962)
3. Wakil, S. J., Pugh, E. L., and Sauer, F., *Proc. Nat. Acad. Sci. U.S.*, **52**, 106 (1964)
4. Goldman, P., *J. Biol. Chem.*, **239**, 3663 (1964)
5. Alberts, A. W., Goldman, P., and Vagelos, P. R., *J. Biol. Chem.*, **238**, 557 (1963)
6. Majerus, P. W., Alberts, A. W., and Vagelos, P. R., *Proc. Nat. Acad. Sci. U.S.*, **51**, 1231 (1964)
7. Majerus, P. W., Alberts, A. W., and Vagelos, P. R., *Proc. Nat. Acad. Sci. U.S.*, **53**, 410 (1965)
8. Majerus, P. W., Alberts, A. W., and Vagelos, P. R., *J. Biol. Chem.*, **240**, 4723 (1965)
9. Pugh, E. L., and Wakil, S. J., *J. Biol. Chem.*, **240**, 4727 (1965)
10. Simoni, R. D., Criddle, R. S., and Stumpf, P. K., *J. Biol. Chem.*, **242**, 573 (1967)

11. Majerus, P. W., *J. Biol. Chem.*, **242,** 2325 (1967)
12. Elovson, J., and Vagelos, P. R., *J. Biol. Chem.*, **243,** 3603 (1968)
13. Powell, G. L., Elovson, J., and Vagelos, P. R., *J. Biol. Chem.*, **244,** 5616 (1969)
14. Lipmann, F., *Science*, **173,** 875 (1971)
15. Alberts, A. W., Majerus, P. W., Talamo, B., and Vagelos, P. R., *Biochemistry,* **3,** 1563 (1964)
16. Joshi, V. C., and Wakil, S. J. *Arch. Biochem. Biophys.*, **143,** 493 (1971)
17. Toomey, R. E., and Wakil, S. J. *J. Biol. Chem.*, **241,** 1159 (1966)
18. Greenspan, M. D., Alberts, A. W., and Vagelos, P. R., *J. Biol. Chem.*, **244,** 6477 (1969)
19. Prescott, D. J., and Vagelos, P. R., *J. Biol. Chem.*, **245,** 5484 (1970)
20. Toomey, R. E., and Wakil, S. J., *Biochim. Biophys. Acta*, **116,** 189 (1966)
21. Schulz, H., and Wakil, S. J., *J. Biol. Chem.*, **246,** 1895 (1971)
22. Majerus, P. W., Alberts, A. W., and Vagelos, P. R., *J. Biol. Chem.*, **240,** 618 (1965)
23. Mizugaki, M., Weeks, G., Toomey, R. E., and Wakil, S. J., *J. Biol. Chem.*, **243,** 3661 (1968)
24. Pugh, E. L., Sauer, F., Waite, M., Toomey, R. E., and Wakil, S. J., *J. Biol. Chem.*, **241,** 2635 (1966)
25. Silbert, D. F., and Vagelos, P. R., *Proc. Nat. Acad. Sci. U.S.*, **58,** 1579 (1967)
26. Norris, A. T., Matsumura, S., and Bloch, K., *J. Biol. Chem.*, **239,** 3653 (1964)
27. Kass, L. R., Brock, D. J. H., and Bloch, K., *J. Biol. Chem.*, **242,** 4418 (1967)
28. Rando, R. R., and Bloch, K., *J. Biol. Chem.*, **243,** 5627 (1968)
29. Brock, D. J. H., Kass, L. R., and Bloch, K., *J. Biol. Chem.*, **242,** 4432 (1967)
30. Helmkamp, G. M., and Bloch, K., *J. Biol. Chem.*, **244,** 6014 (1969)
31. Kass, L. R., and Bloch, K., *Proc. Nat. Acad. Sci. U.S.*, **58,** 1168 (1967)
32. Birge, C. H., Silbert, D. F., and Vagelos, P. R., *Biochem. Biophys. Res. Commun.*, **29,** 808 (1967)
33. Weeks, G., and Wakil, S. J., *J. Biol. Chem.*, **243,** 1180 (1968)
34. White, H. B., Mitsuhasi, O., and Bloch, K., *J. Biol. Chem.*, **246,** 4751 (1971)
35. Weeks, G., and Wakil, S. J., *J. Biol. Chem.*, **245,** 1913 (1970)
36. Barnes, E. M., and Wakil, S. J., *J. Biol. Chem.*, **243,** 2955 (1968)
37. Barnes, E. M., Swindell, A. C., and Wakil, S. J., *J. Biol. Chem.*, **245,** 3122 (1970)
38. Stumpf, P. K., and Barber, G. A., *J. Biol. Chem.*, **227,** 407 (1957)
39. Barron, E. J., Squires, C., and Stumpf, P. K., *J. Biol. Chem.*, **236,** 2610 (1961)
40. Harwood, J. L., and Stumpf, P. K., *Lipids*, **7,** 8 (1972)
41. Overath, P., and Stumpf, P. K., *J. Biol. Chem.*, **239,** 4103 (1964)
42. Brookes, J. L., and Stumpf, P. K., *Biochim. Biophys. Acta*, **98,** 213 (1965)
43. Brookes, J. L., and Stumpf, P. K., *Arch. Biochem. Biophys.*, **116,** 108 (1966)
44. Stumpf, P. K., and James, A. T., *Biochim. Biophys. Acta*, **70,** 20 (1963)
45. Devor, K. A., and Mudd, J. B., *Plant Physiol.*, **43,** 853 (1968)
46. Weaire, P. J., and Kekwick, R. G. O., *Biochem. J.*, **119,** 48 p (1970)
47. Huang, K. P., and Stumpf, P. K., *Arch. Biochem. Biophys.*, **143,** 412 (1971)
48. Brindley, D. N., Matsumura, S., and Bloch, K., *Nature (London)*, **224,** 666 (1969)
49. Ilton, M., Jevans, A. W., McCarthy, E. D., Vance, D., White, H. B., and Bloch, K., *Proc. Nat. Acad. Sci. U.S.*, **68,** 87 (1971)
50. Delo, J., Ernst-Fonberg, M. L., and Bloch, K., *Arch. Biochem. Biophys.*, **143,** 384 (1971)
51. Ernst-Fonberg, M. L., and Bloch, K., *Arch. Biochem. Biophys.*, **143,** 392 (1971)
52. Matsumura, S., and Stumpf, P. K., *Arch. Biochem. Biophys.*, **125,** 932 (1968)
53. Ailhaud, G. P., Vagelos, P. R., and Goldfine, H. *J. Biol. Chem.*, **242,** 4459 (1967)
54. Willecke, K., Ritter, E., and Lynen, F., *Eur. J. Biochem.*, **8,** 503 (1969)

55. Hancock, W. S., Prescott, D. J., Nulty, W. L., Weintraub, J., Vagelos, P. R., and Marshall, G. R., *J. Amer. Chem. Soc.*, **93**, 1799 (1971)
56. Lynen, F., Oesterhelt, D., Schweizer, E., and Willecke, K. In *Cellular Compartmentalization and Control of Fatty Acid Metabolism*, p. 1. Ed. by Gran, F. C. Universitetsforlaget. Academic Press Inc., London and New York, 1968
57. Reed, L. J., and Cox, D. J., *Annu. Rev. Biochem.* **35**, 57 (1966)
58. Van den Bosch, H., Williamson, J. R., and Vagelos, P. R., *Nature* (*London*), **228**, 338 (1970)
59. Croonan, J. E., *Biochim. Biophys. Acta*, **144**, 695 (1967)
60. Scheuerbrandt, G., and Bloch, K., *J. Biol. Chem.*, **237**, 2064 (1962)
61. Lennarz, W. J., Scheuerbrandt, G., and Bloch, K., *J. Biol. Chem.*, **237**, 664 (1962)
62. Fulco, A. J., and Bloch, K., *J. Biol. Chem.*, **239**, 993 (1964)
63. Fulco, A. J., Levy, R., and Bloch, K., *J. Biol. Chem.*, **239**, 998 (1964)
64. Schroepfer, G. J., and Bloch, K., *J. Biol. Chem.*, **240**, 54 (1965)
65. Fulco, A. J., *J. Biol. Chem.*, **245**, 2985 (1970)
66. Mudd, J. B., and Stumpf, P. K., *J. Biol. Chem.*, **236**, 2602 (1961)
67. Nagai, J., and Bloch, K., *J. Biol. Chem.*, **240**, PC3702 (1965)
68. Nagai, J., and Bloch, K., *J. Biol. Chem.*, **243**, 4626 (1968)
69. Nagai, J., and Bloch, K., *J. Biol. Chem.*, **241**, 1925 (1966)
70. Harris, R. V., and James, A. T., *Biochim. Biophys. Acta*, **106**, 456 (1965)
71. Morris, L. J., Harris, R. V., Kelly, W., and James, A. T., *Biochem. J.*, **109**, 673 (1968)
72. Vijay, I. K., and Stumpf, P. K., *J. Biol. Chem.*, **246**, 2910 (1971)
73. Hoffman, K., and Lucas, R. A., *J. Amer. Chem. Soc.*, **72**, 4328 (1950)
74. Kaneshiro, T., and Marr, A. G., *J. Biol. Chem.*, **236**, 2615 (1961)
75. O'Leary, W. M., *J. Bacteriol.*, **77**, 367 (1959); **78**, 709 (1959)
76. O'Leary, W. M., *J. Bacteriol.*, **84**, 967 (1962)
77. Zalkin, H., Law, J. H., and Goldfine, H., *J. Biol. Chem.*, **238**, 1242 (1963)
78. Chung, A., and Law, J. H., *Biochemistry*, **3**, 967 (1964)
79. Thomas, P. J., and Law, J. H., *J. Biol. Chem.*, **241**, 5013 (1966)
80. Pohl, S., Law, J. H., and Ryhage, R., *Biochim. Biophys. Acta*, **70**, 583 (1963)
81. Akhtar, M., Hunt, P. F., and Parvez, M. A., *Biochem. J.*, **103**, 616 (1967)
82. Akamatsu, Y., and Law, J. H., *J. Biol. Chem.*, **245**, 701 (1970)
83. Lederer, E., *Biochem. J.*, **93**, 449 (1964)
84. Jauréguiberry, G., Law, J. H., McClosky, J. A., and Lederer, E., *Biochemistry*, **4**, 347 (1965)
85. Gastambide-Odier, M., Delau;meny, J. M., and Lederer, E., *Biochim. Biophys. Acta*, **70**, 670 (1963)

CHAPTER 4

Control Mechanisms Involved in Fatty Acid Synthesis

It is a commonly experienced phenomenon that when an individual's food intake exceeds his metabolic requirements, he tends to put on weight. A limited proportion of this excess carbohydrate is stored as glycogen but the major portion is initially converted into fatty acids followed by their subsequent esterification into triglyceride (Chapter 5). Similarly, under conditions of impaired glucose metabolism such as those associated with starvation or diabetes, and also with dietary regimens high in fat content, fatty acid synthesis is greatly depressed. A parallel increase in the rate of β-oxidation occurs in this situation that has the effect of sparing glycolysis and glucose oxidation *via* the tricarboxylic acid cycle. Indeed, high rates of fatty acid oxidation tend to stimulate gluconeogenesis. This directing influence is favoured by the specific activating effect of acetyl-CoA on pyruvate carboxylase [pyruvate: carbon dioxide ligase (ADP), EC 6.4.1.1] and its inhibitory action on pyruvate dehydrogenase [pyruvate: lipoate oxidoreductase (acceptor-acetylating) EC 1.2.4.1], thereby driving pyruvate away from the reactions of the tricarboxylic acid cycle. Restoration of a normal diet or one high in carbohydrate after a period of fasting results in the usual levels of fatty acid synthesis. Indeed, a strong correlation exists between the activity of certain enzymes and the rate of *de novo* fatty acid synthesis with nutritional status. Evidently, therefore, control mechanisms are of considerable importance in switching on or off the process of fatty acid synthesis and these will be discussed shortly. However, the intracellular distribution of the substrates and the role of the enzymes involved in translocating the acetyl group to the correct site in the cytoplasm must first be considered. The localization of enzymes concerned with citrate metabolism has recently been covered in depth by Greville[1] in an article that should be consulted for further information.

ORIGIN OF ACETYL-CoA IN THE CYTOPLASM

After degradation the carbon chain of most food-stuffs eventually forms acetyl-CoA within the mitochondria (Scheme 4.1). It has been well documented in the preceding chapters that fatty acid synthesis in animals, fungi and bacteria is mediated by enzymes that are found in the soluble (non-particulate) portion of the cytoplasm. Acetyl-CoA is formed intramito-

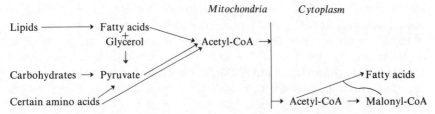

Scheme 4.1   Role of acetyl-CoA in fatty acid biosynthesis

chondrially mainly by the action of pyruvate dehydrogenase and the enzymes of β-oxidation, and the acetyl residue must therefore be transferred in some way across the mitochondrial membrane (Scheme 4.1). A number of possible mechanisms may be envisaged. They have been considered in detail by Lowenstein[2] but most have been discarded on various grounds.

The simplest mode of entry of acetyl-CoA into the cytoplasm would be by direct transfer[2] (Scheme 4.2). However, this process of diffusion is exceedingly

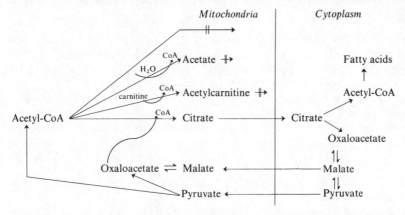

Scheme 4.2   Possible means of transfer of the acetyl groups of acetyl-CoA from the mitochondria to the cytoplasm

slow (since intact mitochondria are impermeable to this substrate) compared with the observed rate of synthesis of fatty acids *in vivo* or even in tissue slices. It was therefore proposed that the acetyl group might leave the mito-chondria after hydrolysis to acetate[3] and be regenerated in the cytoplasm by reaction with acetyl-CoA synthetase [acetate:CoA ligase (AMP), EC 6.2.1.1]. This enzyme does indeed occur in the cytoplasm but the intramito-chondrial level of acetyl-CoA hydrolase (EC 3.1.2.1) is low. With regard to the possible intermediacy of acetyl carnitine, Bremer[4] and Fritz[5] have both suggested that acetyl-CoA:carnitine O-acetyltransferase (EC 2.3.1.7) may be involved in the utilization of acetyl groups for fatty acid synthesis. However, although this enzyme is present in rat liver mitochondria it is absent from the extramitochondrial region of the cell. Incorporation of acetyl carnitine into fatty acids is considerably slower than the incorporation of acetate or citrate. Further, the level of the transferase is much greater in rat heart and skeletal muscle than liver,[6] a distribution that is inconsistent with a possible role in fatty acid synthesis. Carnitine seems to play an important part in the metabolism of long-chain fatty acids and is related to the transfer of these acyl groups across the inner mitochondrial membrane, which is impermeable to coenzyme A derivatives, to the site of β-oxidation.

Thus the above possibilities may be discounted as important routes for entry of acetyl-CoA into the cytoplasm. Considering the final alternative, the pathway in which acetyl-CoA leaves the mitochondria as carbon atoms 1 and 2 of citrate may therefore operate. Lowenstein[2] has proposed that citrate is formed intramitochondrially from acetyl-CoA and oxaloacetate by the action of citrate synthase [citrate oxaloacetate lyase (CoA acetylating), EC 4.1.3.7] (equation (1)):

$$\text{Acetyl-CoA} + \text{oxaloacetate}^{2-} + H_2O \rightarrow \text{citrate}^{3-} + \text{CoASH} + H^+ \tag{1}$$

The reaction is highly exergonic (-9.1 kcal/mole) and is essentially irreversible. The mitochondrial membrane is accessible to citrate which then diffuses out into the cytoplasm where it may be cleaved by the action of ATP citrate lyase [ATP:citrate oxaloacetate-lyase (CoA acetylating and ATP dephosphorylating) (EC 4.1.3.8)][7,8] to regenerate acetyl-CoA and oxaloacetate extramitochondrially (equation (2)):

$$\text{Citrate}^{3-} + \text{ATP}^{3-} + \text{CoASH} \rightarrow \text{acetyl-CoA} + \text{oxaloacetate}^{2-} + \text{ADP}^{2-} + \text{HPO}_4^{2-} \tag{2}$$

This mechanism for the utilization of citrate is supported by the fact that citrate lyase activity varies in the expected manner according to hormonal and nutritional states.

Spencer and Lowenstein[9,10] demonstrated that mitochondria were in fact capable of synthesis and excretion of citrate at a rate that is consistent with observed rates of fatty acid synthesis. Movement of citrate into the cytoplasm or mitochondria was dependent on a malate transporting system[11] which required the presence of malate and orthophosphate. Addition of n-butylmalonate, an inhibitor of malate permease, decreased fatty acid synthesis in a mitochondria plus supernatant system from alanine but not from citrate.[12] Formation from alanine was dependent on the intermediate production of acetyl-CoA and its removal from the mitochondria.

Lowenstein[9,10] and Srere[13] with their coworkers showed that citrate was incorporated more rapidly than acetate into fatty acids and that the acetyl portion was involved in this process. An even greater rate was noted for liver extracts from starved animals that had been refed on a high carbohydrate diet (Table 4.1). The incubation mixture contained the same

Table 4.1 Incorporation of $[1,5-^{14}C]$citrate and $[2-^3H]$acetate into fatty acids by a high-speed supernatant fraction of rat liver (from Spencer et al., 1964[10] and Lowenstein, 1968)[2]

Nutritional state	Precursor incorporated (μm moles/mg protein/hr)		Ratio of citrate/acetate incorporated	Citrate plus acetate incorporated (μm moles/g tissue /hr)
	Citrate	Acetate		
Normal diet	15·0	1·4	11	1,700
Starved for 24 hr	4·9	0·6	8	565
Starved for 48 hr	0·8	0·8	1	165
Starved for 48 hr; refed for 52–60 hr	105·0	1·5	70	11,090

concentration of $[1,5-^{14}C]$citrate and $[2-^3H]$acetate as labelled substrates. Moreover, fatty acid synthesis from acetate was not increased after starving and refeeding. A similar but smaller increase in the ratio of citrate to acetate incorporation was also found for rat mammary gland. Avidin abolished the incorporation of citrate into fatty acids,[13] showing that the conversion was dependent on the carboxylation of acetyl-CoA, that is on the intermediacy of malonyl-CoA.

Thus the first stage in fatty acid synthesis in animal tissues may be considered as the intramitochondrial condensation of acetyl-CoA with oxaloacetate to give citrate. The citrate thus formed may undergo the reactions of the tricarboxylic acid cycle or act as the precursor of acetyl-CoA and initiate lipogenesis.[14] In this case, it diffuses out into the cytoplasm where it is subjected to the action of citrate lyase. The oxaloacetate that is formed cannot readily cross the mitochondrial membrane[15,16] and is reduced by the cytoplasmic malate dehydrogenase[17] (EC 1.1.1.37) to malate (equation (3)):

$$\text{Oxaloacetate} + \text{NADH} + \text{H}^+ \rightleftharpoons \text{malate} + \text{NAD}^+ \tag{3}$$

This may return to the mitochondria where it is reoxidized to oxaloacetate.[18] Alternatively, malate may be converted into pyruvate by malic enzyme [L-malate:NADP oxidoreductase (decarboxylating) (EC 1.1.1.40)] with the generation of NADPH and CO_2 (Scheme 4.2) (equation (4)):

$$\text{Malate} + \text{NADP}^+ \rightleftharpoons \text{pyruvate} + \text{NADPH} + \text{H}^+ + \text{CO}_2 \tag{4}$$

This enzyme is found in liver[14,19] and adipose tissue[20] and is localized exclusively in the supernatant fraction. It is concerned with the production of some of the reducing equivalents utilized for fatty acid synthesis. The pyruvate formed by the oxidative decarboxylation of malate enters the mitochondria where it may be oxidized to acetyl-CoA by pyruvate dehydrogenase or carboxylated to give oxaloacetate and hence restore the level of C_4

dicarboxylic acids.[21] Pyruvate carboxylase performs the latter reaction for which there is an absolute requirement for acetyl-CoA. This sequence (Scheme 4.2) has received further support during studies on acetyl-CoA and NADPH utilization in adipose tissue.[20]

CONTROL OF ENZYMES INVOLVED IN FATTY ACID SYNTHESIS

Citrate Synthase

A number of factors are known to control the activity of citrate synthase from various sources. ATP at physiological concentration inhibits the enzymes isolated from yeast and pig heart by markedly reducing their affinity for acetyl-CoA.[22] ADP and AMP also inhibit but they are less effective. The net kinetic response to the [ATP]/[ADP]/[AMP] balance is more critical in controlling this inhibition. Shepherd and Garland[23] later confirmed this effect with a purified enzyme from rat liver mitochondria. The relationship between inhibition and [ATP] gave rise to sigmoid kinetics; inhibition was non-competitive with respect to acetyl-CoA or oxaloacetate.

It is well documented that conditions which give rise to a high rate of fatty acid oxidation (starvation, diabetes, diet high in fat) result in the mitochondrial formation of ketone bodies. Wieland and coworkers[24] have argued that the production of acetoacetate under these conditions is caused by a fall in the mitochondrial concentration of oxaloacetate after reduction and a consequent direction of acetyl-CoA away from the tricarboxylic acid cycle towards acetoacetyl-CoA formation by reaction with acetoacetyl-CoA thiolase (EC 2.3.1.9). Ketosis is therefore at least partially controlled by the high level of NADH that is produced during fatty acid oxidation with the resultant formation of malate (equation (3)) which then leaves the mitochondria. In the cytoplasm, malate is reconverted to oxaloacetate by an isoenzyme of malate dehydrogenase, generating NADH in the process. Both products are then utilized for glucose synthesis. It has been calculated[25] from the mitochondrial [NAD$^+$]/[NADH] ratio in rat liver that the oxaloacetate concentration would fall below the K_m value for citrate synthase. Thus if the [NAD$^+$]/[NADH] ratio falls, [oxaloacetate]/[malate] would also fall. If the sum of [oxaloacetate] plus [malate] remains steady, it follows that [oxaloacetate] must fall. The direct inhibition of citrate synthase by ATP may also be considered as regulatory.

A further possible source of inhibition of citrate synthase may be long-chain acyl-CoA thioesters. Many workers[26,27] have shown that palmitoyl- and stearoyl-CoA strongly inhibit citrate synthase *in vitro*. Indeed the concentration of long-chain acyl-CoA esters does fluctuate greatly with nutritional status in a direction that is consistent with their fulfilling a regulatory role; it is elevated in the liver[28,29] under conditions of increased fatty acid oxidation or reduced synthesis. It was therefore proposed that they might act as an inhibitory control (negative modifier) for citrate synthase and hence limit the production of extramitochondrial acetyl-CoA. However, the extent of inhibition is dependent on the relative palmitoyl-CoA and enzyme concentration and it may be prevented or even reversed under experimental conditions if sufficient protein (serum albumin) is present.

This inhibition is relatively non-specific and is shown by many other enzymes including acetyl-CoA carboxylase[30] [acetyl-CoA:carbon dioxide ligase (ADP), EC 6.4.1.2] and fatty acid synthetase.[28] The proposition then emerged that these metabolites may regulate fatty acid synthesis *in vivo*. However, Taketa and Pogel studied the effect of palmitoyl-CoA on a series of 12 enzymes, many of which were related to fatty acid synthesis but including others of quite separate metabolic function.[31] Palmitoyl-CoA did indeed inhibit glucose 6-phosphate dehydrogenase competitively and also 6-phosphogluconate dehydrogenase; these enzymes are concerned with the cytoplasmic generation of NADPH for fatty acid and other syntheses. However, malic enzyme (which serves the same function) was not inhibited even at a level 100-fold greater than that required to inactivate glucose 6-phosphate dehydrogenase. Moreover, glutamate dehydrogenase, an enzyme unrelated to fatty acid synthesis, was the most sensitive of those tested. Thus the widespread nature of this inhibition by palmitoyl-CoA seems to preclude any specific physiological role as a regulator of lipogenesis. More probably, the inhibitory effects may be related to the detergent-like properties of long-chain acyl-CoA thioesters that enable them to bind to proteins. This could cause general conformational changes in many enzymes and affect their apparent activity.

ATP Citrate Lyase

The level of citrate lyase is very susceptible to a number of nutritional conditions that control fatty acid metabolism. Thus its catalytic activity is strongly suppressed in rat liver during starvation but is rapidly restored on feeding with a high carbohydrate diet.[3,32] Lowenstein[2] and others[33,34] contended that a reduction in fatty acid synthesis might be due to decreased citrate lyase activity. Addition of large amounts of glucose, fructose or glycerol to the normal animal diet caused great increases in citrate lyase activity over the starvation level, compared with a somewhat smaller increase on refeeding the standard diet. Similar additions to animals on normal, low-fat diets without prior starvation also gave rise to a much enhanced activity. These results complement and confirm those previously obtained by Spencer *et al.*[10] on the overall effect on fatty acid synthesis in starved animals after refeeding. Moreover, a high glucose diet had no effect on the level of citrate lyase in rats that had been made diabetic by treatment with alloxan. This procedure gives rise to a diabetic condition that is mediated through an effect on the insulin-producing β-cells in the islets of Langerhans in the pancreas.[35] Under these circumstances and in starvation, fatty acid synthesis and citrate lyase activity were also reduced in liver[36] (and adipose tissue[33]) but injection of insulin to diabetic rats restored the level of these activities to normal.[32]

Citrate lyase activity (together with that of acetyl-CoA carboxylase) in rat mammary gland also increases greatly at the onset of lactation but falls rapidly after weaning. There is a parallel upsurge and decline in fatty acid synthesis at these times.[34,8] The activity of citrate lyase in livers of mice that carry a recessive gene for obesity has also been compared with that found in normal non-obese mice.[8] When the obesity became apparent, citrate lyase increased greatly compared with the controls. Acetokinase was also examined for comparison but its activity did not vary in the two types, again implying that this enzyme did not play a significant role in the formation of cytoplasmic acetyl-CoA.

Adipose tissue is an active site of lipogenesis and its level of citrate lyase has also been followed with respect to diet.[18] The tissue was homogenized and centrifuged after which the clear supernatant (not containing the upper fat layer) was removed. The level of activity of citrate lyase, based on tissue nitrogen, exceeded that of other tissues and this was greatly increased after fasting and refeeding with a high carbohydrate–low fat diet. In addition, a similar rise in the level of malic enzyme occurred. With regard to the consequence of insulin deficiency, citrate lyase activity was also markedly reduced in adipose tissue of alloxan-treated rats.[33]

More recent evidence implicating ATP citrate lyase as an important factor has since been obtained.[12] Hydroxycitrate, an inhibitor of this enzyme, greatly reduced fatty acid synthesis from alanine (and hence pyruvate and acetyl-CoA) in a supernatant plus mitochondrial system under conditions in which other metabolic functions such as transaminase or pyruvate dehydrogenase activity were not affected.

Foster and Srere[37] have further tested the relationship between fatty acid synthesis and citrate lyase activity. With fasting rats, fatty acid synthetase activity did actually fall sharply in liver slices but citrate lyase activity remained normal for some considerable time. When the test animals were refed with a high carbohydrate diet, fatty acid synthesis increased rapidly but initially at a rate that was not paralleled by citrate lyase. Fatty acid synthetase activity increased nearly 30-fold over the control fasted level, whereas citrate lyase remained unchanged during this early period. Administration of alloxan decreased fatty acid synthesis before any change was noted for citrate lyase levels (a similar situation with that in fasting conditions), but both activities fell with time. Addition of purified citrate lyase to fatty acid synthetase preparations from livers of fed or fasted animals did not result in any increase in fatty acid synthesis. Thus lack of citrate lyase did not generate any immediate limitation over this process.

The block in fatty acid synthesis that occurs in the livers of fasted or diabetic animals therefore appears to lie after the production of acetyl-CoA, at the level of acetyl-CoA carboxylase or fatty acid synthetase. Thus regulation of citrate lyase may not be responsible for the immediate control of fatty acid synthesis in liver. However, the adaptive changes representing stimulation or inhibition that have been reported in various tissues by Lowenstein[3,10,32] and others[18,33,34] for this enzyme under numerous conditions may be related to the operation of a long-term regulatory mechanism for fatty acid synthesis. The lower levels of activity may reflect a repression of *de novo* synthesis of enzyme. Parallel changes in the activity of malate dehydrogenase (decarboxylating) (EC 1.1.1.40) that correlate with the nutritional or hormonal status have also been found.

Citrate lyase is inhibited by ADP, a product of the reaction it catalyses (equation (2)). This reaction shows typical Michaelis kinetics of rate against [ATP] under conditions of varying [ADP].[38] The apparent K_m for ATP is increased and the inhibition is competitive. Thus this enzyme is active in situations in which the ATP level is high.

Acetyl-CoA Carboxylase

The carboxylation to give malonyl-CoA has often been considered the rate-limiting step in the overall process of fatty acid synthesis. Moreover, acetyl-CoA itself is positioned at a number of major cross-roads in metabolism for both catabolic and anabolic reaction sequences (Chapter 1, Scheme 1.1). Acetyl-CoA carboxylase is therefore ideally suited to play a key regulatory role and a great deal of intensive effort has been expended in studying this enzyme from different tissues under various conditions. In the course of the first experiments with cell-free systems from pigeon liver, Brady and Gurin[39] reported that the incorporation of [14C]acetate into fatty acids was greatly enhanced in the presence of citrate. Many workers later confirmed this stimulatory effect with citrate or isocitrate in a number of animal tissues. It could not be explained simply by the production of additional NADPH or bicarbonate in the course of the isocitrate dehydrogenase reaction. Popják and Tietz[40] similarly reported that malonate (especially in the presence of α-oxoglutarate) strongly stimulated fatty acid synthesis from acetate in the mammary gland. Accordingly it was concluded that these acids possessed a further role and it was later demonstrated that the site of activation was acetyl-CoA carboxylase. Experiments with a purified preparation of this enzyme from adipose tissue of rat[41,42] established that the mechanism of activation was not associated with a transcarboxylation reaction. In the absence of citrate, fatty acid synthetase activity was apparently very much greater than that of acetyl-CoA carboxylase, indicating that this latter reaction exerted a rate-limiting function. Addition of citrate sharply increased the rate of the carboxylase but did not affect the synthetase.

Lynen and his colleagues[43] examined the effect of citrate and isocitrate in rat liver on the individual reaction steps catalysed by acetyl-CoA carboxylase, that is, the carboxylation of biotinyl enzyme involving the hydrolysis of ATP and the subsequent transcarboxylation to the substrate. Both reactions were greatly stimulated.

Acetyl-CoA carboxylase may be assayed routinely by the $H^{14}CO_3^-$ fixation method. Acetyl-CoA is incubated with enzyme and $H^{14}CO_3^-$ and the reaction is terminated after a suitable period by the addition of acid. The principle behind this procedure lies in the fact that the product, malonyl-CoA, is stable under these conditions but the unchanged bicarbonate is released as $^{14}CO_2$. Thus the incorporation of radioactivity which is equivalent to the activity of enzyme is simply proportional to the amount of acid-stable radioactivity in the reaction mixture. In all cases, the enzyme has been examined in livers from animals that have been starved for 2 days and then refed with a diet high in carbohydrate and low in fat for 3–4 days.

Vagelos and coworkers[42,44] later showed that the activating effect on the adipose tissue carboxylase was associated with a change in its sedimentation pattern and therefore related to a modification in the conformation of the protein. Acetyl-CoA carboxylase sedimented on sucrose gradient centrifugation with a sedimentation ($s_{20,w}$) coefficient of 19S but on incubation with citrate this value increased to approximately 43S (that is, it sedimented at a much faster rate). Thus activation of the enzyme was accompanied by an aggregation of the protein to give a trimer product. Similar effects were noted with other activators. The temperature of incubation with citrate was critical since activation and aggregation did not occur at $0°$.

Meanwhile, Waite and Wakil[45] had also demonstrated that acetyl-CoA carboxylase from chicken liver was strongly stimulated by a number of tricarboxylic and dicarboxylic acids with isocitrate as the most effective. Lane and coworkers[46,47] continued these studies with the enzyme purified to homogeneity and confirmed that it was strikingly activated (15-fold) by citrate or isocitrate and also by malonate. The activation, unlike that in rat liver and adipose tissue, did not require preincubation with the effector but was also accompanied by an increase in the sedimentation coefficient of the enzyme in sucrose density gradients. The inactive protomeric form ($s_{20,w}$ value 13S; molecular weight 410,000) was converted into an active polymeric filamentous form (10×400 nm) as evidenced by electron microscopy, with a high molecular weight ranging from 4–8×10^6 (56–59S) under various conditions of buffer and pH.[48] The minimum length required for activity has not yet been determined. There is one biotin prosthetic group and one binding site for HCO_3^-, acetyl-CoA and citrate per mole of protomer. The enzyme is particularly stable at room temperature in the presence of activators (isocitrate, citrate and phosphate) because of the filamentous nature of the polymeric form of the enzyme. The activation results from an increase in V_{max} of the carboxylation reaction.

The sedimentation coefficient of a highly purified enzyme from chicken liver was also measured by analytical ultracentrifugation[49] and its value determined as 14S and 45S in the absence and presence of citrate respectively. The former value especially was in good agreement with that determined by Lane's group.[48] Similar data were obtained by a sucrose density gradient method. The enzyme has since been crystallized in a rod-like form.[50]

Lynen and colleagues have reported similar results with a purified enzyme from rat liver. An increase in the sedimentation coefficient coupled with activation of the enzyme was caused by incubation with citrate.[51] Kinetic studies[52] demonstrated that this activation was also caused by an increase in V_{max} for the reaction. However, addition of palmitoyl-CoA prevented this rise in the sedimentation coefficient. Acetyl-CoA carboxylase was also inhibited by long-chain acyl-CoA compounds in a non-competitive manner with respect to its substrates but competitively with regard to citrate. The reduced activity of this enzyme, apparent during starvation or diabetes, might therefore be due to inhibition by such compounds since they accumulate under these conditions.[52] However, it was possible that the low levels of acetyl-CoA carboxylase might be due, at least partially, to repression at the enzyme synthesis level caused by the presence of these metabolites. It must be emphasized that the regulatory aspects of these results should be treated with caution since it has been established that the inhibitory effects of palmitoyl-CoA is widespread amongst enzymes and may often be prevented at high-protein concentration.[31] In this case, however, the concentration of the acyl-CoA thioester required was very low at 1 μM.

Lowenstein and colleagues[29,53] also worked with acetyl-CoA carboxylase from rat liver and confirmed that it required prior incubation with citrate or Mg^{2+} before activation became apparent. Even in the presence of Mg^{2+}, citrate was still required for maximum stimulation. Addition of ATP, a substrate of the reaction, initially increased the activity but excess reversed the activating effect of both citrate and Mg^{2+} and caused considerable inhibition. Activation with citrate resulted in an increase in sedimentation coefficient (conversion into a 'heavy' form) but activation with Mg^{2+} caused it to sediment as the normal 'light' form.

Addition of Mg^{2+} also gave partial activation to the enzyme from the mammary gland of lactating rats but citrate was again necessary for full stimulation.[54] Increasing the citrate concentration gave rise to a sigmoid curve for the kinetics of the incorporation of HCO_3^- by this enzyme, suggesting that a cooperative (allosteric) interaction might be occurring. This has proved a particularly attractive proposition since a high citrate concentration would seem to predispose favourable conditions for fatty acid synthesis.

Doubts have frequenty been expressed, however, concerning the validity of a physiological role for citrate as an activator of acetyl-CoA carboxylase. It is indeed difficult to rationalize this purported role since its extramitochondrial (or even cellular) concentration is apparently much lower than the 4–7 mM required for half-maximum stimulation in rat liver and, in any event, it does not change appreciably under various conditions.[29,51] The liver content remains remarkably constant at approximately 0·6 mM during starvation, after administration of alloxan or even feeding with a high carbohydrate diet.[55] Moreover a rather different situation applies in adipose tissue where citrate concentration is increased after treatment with adrenaline or alloxan (conditions associated with a depression of fatty acid synthesis).[56] A further study[57] confirmed that there was no correlation between citrate concentration and fatty acid synthesis after administration of insulin. An interpretation that would reconcile these experimental results is that acetyl-CoA carboxylase may be further modified by another metabolite that lowers the concentration of citrate required to nearer the cellular level.

It is possible, however, that the citrate concentration at the site of the carboxylase may be greater than that expected from the intracellular value if the enzyme is associated with some membranous structure *in vivo*.[53] Acetyl-CoA carboxylase does in fact occur in both microsome and supernatant fractions in certain animal tissues but this may simply be related to the isolation procedure; the pigeon liver[58] and rabbit mammary gland[59] enzymes, for instance, are probably membrane-bound *in vivo*. Moreover, the carboxylase remains tightly bound to microsomes in sonicated yeast extracts.[60] An interesting study by Iliffe and Myant[61] recently revealed that citrate did not activate this enzyme in liver homogenates or purified preparations that had been supplemented with microsomes (Table 4.2). In these systems the activity of acetyl-CoA carboxylase was the rate-limiting step. Indeed, it has also been shown that this enzyme may be activated (to the same extent as citrate) simply by addition of phospholipids derived from membranes of the endoplasmic reticulum.[62] Iliffe and Myant[61] have advanced the view that this enzyme becomes insensitive to citrate when associated with membranous particles (and similarly in the intact liver cell), by virtue of taking up a favourable conformation or, alternatively, because inhibitory end-products such as acyl-CoA esters are rapidly removed under these conditions by conversion into lipids. In this situation, its activity is already expressed with maximum efficiency. These workers[61] have therefore argued that an active membrane–carboxylase complex may exist *in vivo* to which the fatty acid synthetase (present in the cell-sap) may be loosely attached. However, there is no indication that the yeast synthetase is bound to membranous structures or contains any lipid.[63] Despite this it seems logical to suppose that there would be a loose binding between the synthetase

Table 4.2   Effect of citrate on the incorporation of [1-^{14}C]acetate into fatty acids by various subcellular preparations from rat liver (from Iliffe and Myant, 1970)[61]

In these preparations, the activity of acetyl-CoA carboxylase was rate-limiting compared with that of acetyl-CoA synthetase and fatty acid synthetase

| | Concentration of citrate (mM) | | [^{14}C]Acetate incorporated (nmol/h/mg of protein) |
Preparation	Added	Final	
A. Homogenate	0	< 0·2	31
	20	7·6	20
B. Post-mitochondrial	0	<0·2	6
supernatant	20	12·7	11
C. Post-microsomal	0	<0·2	0·2
supernatant	20	14·5	11
D. As (C) plus	0	0·7	9
microsomes	20	13·0	12

and endoplasmic reticulum *in vivo* that permits ready transfer of acyl groups to lipid precursors.

Finally, it should be pointed out that citrate specifically inhibits phosphofructokinase (ATP-D-fructose 6-phosphate 1-phosphotransferase, EC 2.7.1. 11)[64] and therefore limits the supply of fructose-1,6-diphosphate and hence acetyl-CoA (and glycerophosphate) for fatty acid and lipid synthesis. Moreover, a high ATP concentration reverses the apparent activation of acetyl-CoA carboxylase by citrate. The consequences of these important metabolic interactions would obviously be paradoxical *in vivo* since ATP and a rapid glycolytic flux are a prerequisite for lipogenesis.

It therefore seems more probable that changes in acetyl-CoA carboxylase activity following alterations in dietary status may be due to an adaptive phenomenon (that is, an effect on enzyme synthesis) rather than an effector-mediated stimulation or inhibition of the enzyme directly.[52] This concept was recently followed up with interesting studies that made use of immunological analysis for assay of this enzyme. Antibodies prepared from chicken[65] or rabbit[66] liver acetyl-CoA carboxylase cross-reacted equally efficiently with enzyme from rat liver and the extent of precipitation was proportional to enzymic activity. They firmly demonstrated that the rise and fall in activity associated with dietary change was due to differences in the enzyme content itself. Immediate short-term regulation of fatty acid synthesis, therefore, would not be accomplished by this means. The existence of a specific inducer under conditions of carbohydrate excess was also implied but its identity has not yet been established.[65] Moreover these immunochemical studies verified that the increased activity in genetically obese mice was mainly due to a rise in the amount of enzyme protein.[67]

Fatty Acid Synthetase

It has been demonstrated that fatty acid synthetase from rat[25] and pigeon[68] liver are inhibited by palmitoyl-CoA (but not palmitate). The inhibition could be prevented but not reversed by the addition of bovine serum albumin and was non-competitive with respect to malonyl-CoA. Treatment of the enzyme from pigeon liver with palmitoyl-CoA caused a large decrease in its sedimentation coefficient with dissociation into inactive sub-units with half the original molecular weight. These effects are thought to be due to disruption of non-covalent bonds in the synthetase.

Wakil's group[69] also showed that this synthetase was inhibited by malonyl-CoA, a substrate of the enzyme complex; the inhibition was competitive with respect to NADPH and was mediated by a sharp increase in its K_m value. This effect was reversed with simultaneous reduction in the elevated K_m value by fructose-1,6-diphosphate. The experimental data, based on elegant kinetic studies, indicated that substrate inhibition due to excess malonyl-CoA was caused by additional binding to the synthetase at a regulatory site but that affinity for this site was lost in the presence of fructose-1,6-diphosphate.

A rather more pronounced and specific inhibition by long-chain acyl-CoA compounds occurs with the purified yeast fatty acid synthetase. Lust and Lynen[70] made a detailed study of their effects and reported that the inhibition was competitive with respect to malonyl-CoA but non-competitive with regard to the remaining substrates, acetyl-CoA and NADPH. It could be reduced by the addition of bovine serum albumin but even relatively high levels of albumin (2 mg/ml) did not reduce the inhibition by more than approximately 50 per cent. As little as 1–2 μM palmitoyl-CoA was required to cause 50 per cent inhibition even in the presence of 0·5 mg albumin/ml. Free palmitate had no significant effect on enzyme activity even in the absence of albumin. In this particular case, the inhibitory effect appears to be authentic, at least in part, and not due to the general detergent-like nature of the long-chain acyl-CoA thioesters. Moreover, since the inhibition is competitive with respect to malonyl-CoA, they must compete with malonyl-CoA for the 'central' 4′-phosphopantetheine thiol site in the yeast synthetase and thus prevent attachment of substrate or elongation of existing acyl groups already bound to the enzyme.

Formation of Triacetic Acid Lactone (I)

The effect of omission of NADPH from systems catalysed by fatty acid synthetases of animal, yeast and bacterial origin has been studied and as expected synthesis is greatly impaired. In all cases, a C_6-compound, triacetic acid lactone (4-hydroxy-6-methylpyran-2-one; (I)) is formed. This product is, in fact, the simplest of the polyacetate structures from which so many fungal natural products are made.

(I)

Bressler and Wakil[71] had observed that the fatty acid synthetase from pigeon liver, when incubated with acetyl-CoA and malonyl-CoA in the absence of NADPH, produced a substance with a pronounced ultraviolet-absorption spectrum (λ_{max} 275 nm). They postulated that this might be a polyketo intermediate related to fatty acid synthesis; it was the first indication that this type of structure might be formed by a non-fungal source. The product was subsequently identified as triacetic acid lactone (TAL). Its synthesis was effected by a purified synthetase preparation that formed palmitate in the presence of NADPH.[72] The lactone is derived from one molecule of acetyl-CoA (C-6 and methyl C) by condensation with two molecules of malonyl-CoA (C-2–C-5). During various purification procedures, the relative activities for fatty acid and TAL synthesis remained constant. Thus the ability to synthesize TAL seems to be an integral part of the fatty acid synthetase of pigeon liver but this only becomes apparent when the first reductase component cannot function. The mechanism suggested for its synthesis was that acetoacetyl-enzyme formed in the condensation reaction might react with a second malonyl unit, in the absence of NADPH, by means of the same condensing enzyme to form triacetyl-enzyme. It appears that further condensation cannot occur and that the binding thiol site is regenerated by release of the triacetyl group as the stable lactone.

More recently, Lynen's group[73] showed that fatty acid synthesis by the yeast complex was also blocked at the acetoacetyl-enzyme stage in the absence of NADPH. A highly purified enzyme catalysed the formation of TAL in a reaction that they termed a 'derailment' mechanism of fatty acid synthesis. The explanation presented for its formation was essentially similar to that given previously for the pigeon liver synthetase.[72] However, Lynen[73] has postulated further details concerning the mechanism involved: the acetoacetyl residue is transferred to the 'peripheral' site (in the condensing enzyme) and in this position may react with an additional malonyl group bound to the 'central' site to give triacetyl-enzyme. The stabilized lactone is then formed by nucleophilic attack of the enolate anion on the thioester bond of 4'-phosphopantetheine and release from the enzyme complex (Scheme 4.3). Tetraacetic acid lactone, a structure composed of four acetate

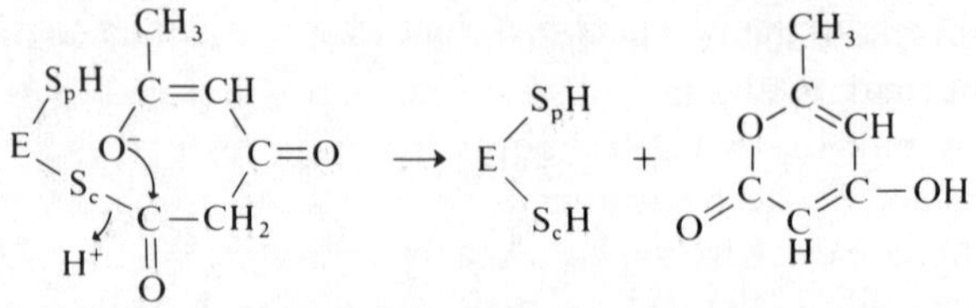

$$\underset{H^+}{\overset{CH_3}{\underset{\displaystyle E{\diagdown}\underset{S_c}{\overset{S_pH}{\diagup}}O}{\underset{\displaystyle \;\;\;C-CH_2}{\overset{C=CH}{\underset{\|}{O}}\diagdown C=O}}}} \;\rightarrow\; E{\diagdown}\overset{S_pH}{\underset{S_cH}{}} \;+\; \underset{O}{\overset{CH_3}{\underset{\displaystyle O{=}\underset{H}{\overset{C}{\diagdown}}}{\underset{C}{\overset{O}{\diagup}}\overset{C}{\underset{\;}{\diagdown}}\overset{CH}{\underset{\displaystyle C-OH}{}}}}}$$

Scheme 4.3   Mechanism for the release of triacetic acid lactone from the yeast fatty acid synthetase (Yalpani et al., 1969)[73]

residues, was not detected among the reaction products, confirming that further condensation did not occur.

The formation of enols as the products of the condensing enzyme has previously been illustrated by Lynen (Chapter 2; Scheme 2.3) without comment; the enolate form of the triacetyl residue is presented here (Scheme 4.3) as the *trans*-isomer. However, if its configuration had been stabilized in a direction away from the thioester bond or had remained as a free carbonyl group, cleavage of this bond would not have taken place at this stage and the potential would still have been available for condensation with yet more malonyl units (Chapter 6). The ability to react in this manner is evidently not inherent in the yeast or pigeon liver synthetases but a similar mechanism might explain the synthesis of polyacetate-derived products in fungi. Presumably the detailed architecture of the 'aromatic' synthetases involved have this essential modification. A detailed discussion of this topic is presented in Chapter 6.

Analogous studies[74] with a preparation of E. coli synthetase also indicated that this enzyme synthesized TAL instead of fatty acids when NADPH was deficient. In this case acetoacetyl-ACP and triacetyl-ACP were proposed as intermediates.

Since the rate of formation of NADPH in the cytoplasm during normal dietary conditions is sufficient to maintain the reductions, there would seem to be little physiological significance in the production of TAL in pigeon liver[72] (or yeast).[73] However, these reactions remain of interest since they throw additional light on the properties of the binding sites within the synthetase with respect to the acyl intermediates.

Enzymes Generating NADPH

Both reductive steps involved in the *de novo* synthesis of fatty acids require NADPH as hydrogen donor. The enzymes utilized for the production of this cofactor all occur in the cytoplasm. In general, it seems that the capacity of the two dehydrogenases of the pentose phosphate pathway, glucose 6-phosphate dehydrogenase (D-glucose 6-phosphate:NADP oxidoreductase, EC 1.1.1.49) and 6-phosphogluconate dehydrogenase [6-phospho-D-gluconate: NADP oxidoreductase (decarboxylating), EC 1.1.1.44], is sufficient to account for normal rates of fatty acid synthesis but they may only contribute approximately half the NADPH required during high rates of lipogenesis in animal tissues.[75] The remainder is formed on conversion of malate to pyruvate by malate dehydrogenase (decarboxylating) (EC 1.1.1.40) (equation (4)). In essence, this is derived from mitochondrial NADH (generated by pyruvate dehydrogenase) after reduction of oxaloacetate to malate and subsequent release into the cytoplasm. Elegant balance studies[76] based on the incorporation of radioactivity into fatty acids in adipose tissue from [2-^{3}H]-glucose and ^{14}C-labelled substrates indicate that, even on a normal diet, a considerable proportion of the reducing equivalents is provided by this means. Both NADPH-generating systems in the liver are depressed during starvation and diabetes but their activity is restored on refeeding or after treatment with insulin and further increased at the onset of lactation.[17,77] Glucose 6-phosphate dehydrogenase is particularly sensitive to insulin and the hepatic activity of this enzyme in fasted or alloxan-treated diabetic rats rises dramatically after feeding or injection of insulin.[78] A similar but smaller effect occurs with 6-phosphogluconate dehydrogenase. Since administration of actinomycin (an inhibitor which interferes with transcription of DNA into RNA and hence prevents protein synthesis) abolishes these increases, insulin presumably acts by stimulating de novo enzyme synthesis. Isocitrate dehydrogenase (EC 1.1.1.42), although present in the cytoplasm and potentially a source of NADPH, remains at a fairly steady level during fasting and refeeding[17] and is also unaffected by diabetes.[79] It does not therefore appear to be involved in generating NADPH for these synthetic reactions.

TERMINATION AT C$_{16}$ AND C$_{18}$ STAGE

In almost all instances the products of de novo fatty acid synthesis are the C$_{16}$ and C$_{18}$ acids and the synthetases concerned obviously possess some inherent mechanism that is responsible for this. An early suggestion by Lynen[80] was that this phenomenon might be related to the specificity of a long-chain acyl-CoA transferase component in the yeast complex but re-examination showed that saturated derivatives throughout the range C$_6$ to C$_{18}$ were transferred at approximately the same rate.[81] The protective effect of acyl-CoA compounds of varying chain-lengths against inhibition by N-ethylmaleimide which binds to both the thiol acceptor sites was then tested with interesting results.[82] Preincubation of yeast synthetase with equimolar concentrations of acetyl-CoA and saturated acyl-CoA homologues protected the thiol groups from inactivation to different extents. Protection was maximal with acetyl-CoA decreasing with increasing chain-length, and therefore seemed related to an ability to acylate the 'peripheral' group within the condensing enzyme. Thus palmitoyl groups were the least efficient of those tested. Malonyl-CoA, which is not attached to this cysteine residue (Chapter 2; Scheme 2.6), provided no protection. Lynen has therefore argued that the environment around this active region is hydrophilic and would tend to repel the long-chain acyl derivatives. Similarly, there might be a lipophilic region around the 'central' thiol group which attracts these derivatives and tends to prevent their transfer to. the active cysteine site in the condensing enzyme. The net consequence of this would be that palmitoyl and stearoyl residues remain attached by thioester linkage to $4'$-phosphopantetheine at the 'central' site and thereby block further condensation with malonyl groups until the fatty acid is liberated (in the case of yeast, by transfer to coenzyme A and in animals, by hydrolysis to the free acid). These studies were extended to measure the inhibitory effect *in vitro* of long-chain acyl-CoA compounds.[70] The inhibition was competitive with respect to malonyl-CoA and therefore confirmed that they competed for the 'central' ACP binding site.

Lynen and coworkers[83] then proposed a modified model to explain chain termination at the level of C_{16} and C_{18} acids. This model assumes that the probability of an enzyme-bound saturated acyl derivative reacting with coenzyme A (and hence terminating the sequence) is related to the relative rates of the condensation (with malonyl-CoA) and transfer reactions. It also supposes that the growing aliphatic chain would not interact with the enzyme until a chain-length of 12–14 carbon atoms is reached. With each additional methylene group added, modification of the relative rates of the two enzymes occurs in favour of transfer. The condensation reaction is rate-limiting with the result that the products of the reaction immediately prior to this, the saturated acyl derivatives, have the greatest probability of becoming the end product. Conditions under which stearoyl-CoA or shorter-chain length derivatives are formed have been calculated and verified since the formula derived by Lynen's group[83] enables the probability of product formation of given chain-length to be determined. Moreover, it indicates the relative concentration of acetyl-CoA and malonyl-CoA required for the synthesis of predominantly long- or short-chain acyl-CoA thioesters. When [malonyl-CoA]/[malonyl-CoA + acetyl-CoA] approaches unity (that is, with trace amounts of acetyl-CoA present), the acids synthesized are predominantly the long-chain C_{16} and C_{18} with minor amounts of C_8–C_{14}. However, when [malonyl-CoA]/[malonyl-CoA + acetyl-CoA] equals 0·1, that is, with the supply of malonyl-CoA restricted (controlled by the amount of acetyl-CoA carboxylase added to the system), the major acid formed is C_{16} with appreciable amounts of C_8–C_{14} and also C_{18}.

Later investigations were aimed at an understanding of the transfer reaction between coenzyme A and saturated acyl derivatives.[81,84] The enzyme responsible was examined using external [^{14}C]coenzyme A and was not sensitive to inhibition by *N*-ethylmaleimide or iodoacetamide indicating that the substrates were bound to non-thiol sites during the course of the reaction. Inhibition of transferase and overall synthetic acitivity was caused, however, at high substrate concentration, as a result of the binding of long-chain acyl-CoA thioesters to the synthetase.[70] Thus if the transferase reaction is slow relative to condensation with malonyl residues but increases in activity with longer chain-length, a preferential release of C_{16} and C_{18} acids would result.

Reaction of [^{14}C]palmitoyl-CoA labelled in the acyl moiety with the yeast synthetase gave rise to a radioactive product.[84] Proteolytic degradation yielded palmitoyl peptides in which the acyl group was bound covalently to the enzyme; as expected some of these proved to be thioesters bound to 4′-phosphopantetheine of ACP or a cysteine residue of the condensing enzyme component. An additional peptide, however, was resistant to treatment with performic acid and contained an oxygen ester grouping which originated from a serine binding site. Its amino acid composition was identical with that determined for the malonyl-transferase peptide.[84] It appears therefore that the malonyl transferase exhibits activity responsible both for elongation (binding of malonyl residues) and termination (reaction of palmitoyl and stearoyl residues with coenzyme A) (Scheme 4.4). This factor may be decisive in influencing the chain-length at which the sequence is stopped. Lack of malonyl-CoA, for instance, results in a 'premature' release of shorter-chain acids[83] in a situation that is favoured by lack of competition between malonyl and long-chain acyl residues at the active serine site of the malonyl transferase enzyme.

Qualitatively similar variations in the nature of the product dependent on [acetyl-CoA]/[malonyl-CoA] had been reported earlier for mammary gland preparations. Addition of a large excess of malonyl-CoA (that is, with acetyl-CoA limiting) resulted in the exclusive production of palmitate and stearate.[85] In contrast, when malonyl-CoA was effectively limiting, approximately equal amounts of C_8–C_{16} acids were formed.[36,87] A direct comparison has since been made[88] between the specificities (with regard to chain termination) of the synthetases from rabbit mammary gland and rat liver, at different malonyl-CoA concentrations. The pattern of acids released from both enzymes was markedly similar with palmitate predominant at high substrate levels. However, as the relative concentration of malonyl-CoA was reduced, synthesis of shorter-chain acids increased to much the same extent in both cases. Thus the specificity for termination at C_{16} is inherent in the two enzymes and it is reasonable to suppose that all animal synthetases may possess similar properties. Synthesis of short- and intermediate-chain acids by purified enzymes or in vivo may simply reflect acetyl-CoA/malonyl-CoA levels (or that of acetyl-CoA carboxylase which would control this balance) in the tissues concerned. However, C_{16} and C_{18} acids are derived additionally from the triglyceride component of circulating lipoproteins.[89]

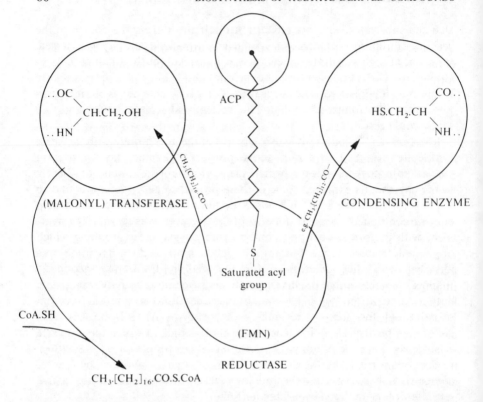

CH$_3$.[CH$_2$]$_{16}$.CO.S.CoA

Scheme 4.4   Partial representation of yeast synthetase showing the reactions engaged by saturated acyl groups. These groups (up to C$_{16}$) may be transferred to the cysteine residue on the condensing enzyme prior to reaction with malonyl-CoA or alternatively (C$_{16}$, C$_{18}$), to the serine residue of the transferase component for reaction with coenzyme A and release from the synthetase. This will result in the liberation of the OH (serine) and SH (4′-phosphopantetheine) sites on the transferase and ACP respectively

By analogy with the yeast system, the components of the animal multi-enzyme complex may also inherently control the release of free fatty acids at the level of C$_{16}$ (and to a lesser extent at C$_{14}$ and C$_{18}$) by competition between the condensation and terminal hydrolase reactions on the one hand and the relative concentration of acetyl-CoA and malonyl-CoA on the other. Furthermore, it is known that pigeon liver synthetase does contain a palmitoyl thioester hydrolase that is specific for the C$_{16}$ and C$_{18}$ saturated[90] or monoenoic acyl-CoA esters but is inactive for myristyl-CoA and thioesters with shorter chain-lengths.[91] However, it should be borne in mind that palmitate is synthesized as the thioester of 4′-phosphopantetheine within the multienzyme complex and presumably therefore this is the actual substrate that is hydrolysed *in vivo*.

CONCLUSIONS

Considerable effort has been expended in attempting to establish which of the individual reactions involved in the conversion of the mitochondrially-formed acetyl-CoA into fatty acids in the cytoplasm are truly regulatory *in vivo*. Much effort, in particular, has centred upon acetyl-CoA carboxylase and the modification of its activity by possible physiological effectors. In addition to the conflicts already raised with regard to the activation by citrate or isocitrate, it has been reported[92] that the capacity of this enzyme may be equivalent to that of fatty acid synthetase and therefore not rate-limiting. Some of the earlier reported values of low carboxylase activity may have been due to suboptimal assay conditions. In view of the multiplicity of reactions catalysed by the multienzyme complex in animals and yeast (or aggregate in bacteria and plants), perhaps this would be a logical site for the fine control that might be dictated by fluctuations in the level of associated metabolites. Possibly, also, regulation may be primarily negative in nature and directed towards prevention of fatty acid synthesis under conditions of impaired glucose utilization. Moreover, studies[93] with intact epididymal fat pads showed that the conversion of glucose into triglyceride was limited by the supply of acetyl-CoA, rather than the activity of any subsequent reaction entailed in fatty acid synthesis. Addition of acetate or pyruvate to the glucose incubation medium increased the rate of fatty acid synthesis in tissue from fed or fasted rats (in the presence or absence of insulin). In adipose tissue under normal conditions, regulation may be accomplished mainly by insulin through controlling the rate of entry of glucose across the cell-membrane.[94] The amount of *de novo* lipogenesis may be limited by the availability of ADP and the ability to utilize the net yield of ATP generated during this process[95] (derived after oxidation of the NADH formed during glycolysis and the pyruvate dehydrogenase reaction).

Some experimental data have been obtained that seem to point in the direction of control at the level of the synthetase, for example, the close relationship between the inhibitory effect of excess malonyl-CoA and its reversal by fructose 1,6-diphosphate (a precursor of acetyl-CoA) in pigeon liver,[69] and the very strong end-product inhibition noted for long-chain acyl-CoA compounds in the yeast system, that is only partially relieved by the addition of albumin.[70] Moreover, the affinity of the malonyl transferase component for its substrate is affected by the prior transfer of acetyl groups to the synthetase.[96] The level of acetyl-CoA carboxylase and the actual concentration of acetyl-CoA and malonyl-CoA available for reaction may control the chain-length of the products released. However, there are indications that the rise in fatty acid synthesis in rat liver on refeeding is associated with an increased incorporation of [14C]leucine into fatty acid synthetase with the inference that this stimulation is at least partially exerted on net enzyme synthesis.[97]

REFERENCES

1. Greville, G. D. In *Citric Acid Cycle*, p. 1. Ed. by Lowenstein, J. M., Marcel Dekker, New York and London, 1969
2. Lowenstein, J. M. In *The Metabolic Roles of Citrate* (Biochemical Society Symposium, No. 27), p. 61. Ed. by Goodwin, T. W. Academic Press Inc., London and New York, 1968
3. Kornacker, M. S., and Lowenstein, J. M., *Biochem. J.*, **94**, 209 (1965)
4. Bremer, J., *J. Biol. Chem.*, **237**, 2228 (1962)
5. Fritz, I. B., *Advan. Lipid Res.*, **1**, 285 (1963)
6. Beenakkers, A. M. T., and Klingenberg, M., *Biochim. Biophys. Acta*, **84**, 205 (1964)
7. Srere, P. A., *J. Biol. Chem.*, **236**, 50 (1961)
8. Spencer, A. F., and Lowenstein, J. M., *Biochem. J.*, **99**, 760 (1966)
9. Spencer, A. F., and Lowenstein, J. M., *J. Biol. Chem.*, **237**, 3640 (1962)
10. Spencer, A. F., Corman, L., and Lowenstein, J. M., *Biochem. J.*, **93**, 378 (1964)
11. Chappell, J. B., *Biochem. J.*, **100**, 43P (1966)
12. Watson, J. A., and Lowenstein, J. M., *J. Biol. Chem.*, **245**, 5993 (1970)
13. Bhaduri, A., and Srere, P. A., *Biochim. Biophys. Acta*, **70**, 221 (1963)
14. Ballard, F. J., and Hanson, R. W., *J. Lipid Res.*, **8**, 73 (1967)
15. Lardy, H. A., Paetkau, V., and Walter, P., *Proc. Nat. Acad. Sci. U.S.*, **53**, 1410 (1965)
16. McLean, P., Brown, J., and Greenbaum, A. L. In *Carbohydrate Metabolism and its Disorders*, Vol. 1, p. 397. Ed. by Dickens. F., Randle, P. J., and Whelan, W. J., Academic Press Inc., London and New York, 1968
17. Pande, S. V., Khan, R. P., and Venkotasubramanian, T. A., *Biochim. Biophys. Acta*, **84**, 239 (1964)
18. Kornacker, M. S., and Ball, E. G., *Proc. Nat. Acad. Sci. U.S.*, **54**, 899 (1965)
19. Wood, H. G., and Utter, M. F. In *Essays in Biochemistry*, Vol. 1, p. 1. Ed. by Campbell, P. N., and Greville, G. D., Academic Press Inc., London and New York, 1965
20. Martin, B. R., and Denton, R. M., *Biochem. J.*, **117**, 861 (1970)
21. Utter, M. F., and Keech, D. B., *J. Biol. Chem.*, **238**, 2603 (1963); Keech, D. B., and Utter, M. F., *J. Biol. Chem.*, **238**, 2609 (1963)
22. Hathaway, J. A., and Atkinson, D. E., *Biochem. Biophys. Res. Commun.* **20**, 661 (1965)
23. Shepherd, D., and Garland, P. B., *Biochim. Biophys. Res. Commun.*, **22**, 89 (1966)
24. Wieland, O., Matschinsky, F., Löffler, G., and Müller, U., *Biochim. Biophys. Acta*, **53**, 412 (1961)
25. Williamson, D. H., Lund, P., and Krebs, H. A., *Biochem. J.*, **103**, 514 (1967)
26. Wieland, O., and Weiss, L., *Biochem. Biophys. Res. Commun.*, **13**, 26 (1963)
27. Srere, P. A., *Biochim. Biophys. Acta*, **106**, 445 (1965)
28. Tubbs, P. K., and Garland, P. B., *Biochem. J.*, **93**, 550 (1964)
29. Fang, M., and Lowenstein, J. M., *Biochem. J.*, **105**, 803 (1967)
30. Bortz, W. M., and Lynen, F., *Biochem. Z.*, **337**, 505 (1963)
31. Taketa, K., and Pogel, B. M., *J. Biol. Chem.*, **241**, 720 (1966)
32. Kornacker, M. S., and Lowenstein, J. M., *Biochem. J.*, **95**, 832 (1965)
33. Brown, J., and McLean, P., *Nature (London)*, **207**, 407 (1965)
34. Howanitz, P. J., and Levy, H. R., *Biochim. Biophys. Acta*, **106**, 430 (1965)
35. Lukens, F. D. W., *Physiol. Rev.*, **28**, 304 (1948)
36. Kornacker, M. S., and Lowenstein, J. M., *Biochim. Biophys. Acta*, **84**, 490 (1964)

37. Foster, D. W., and Srere, P. A., *J. Biol. Chem.*, **243**, 1926 (1968)
38. Atkinson, D. E., and Walton, G. M., *J. Biol. Chem.*, **242**, 3239 (1967)
39. Brady, R. O., and Gurin, S., *J. Biol. Chem.*, **199**, 421 (1952)
40. Popják, G., and Tietz, A., *Biochem. J.*, **60**, 147 (1955)
41. Martin, D. B., and Vagelos, P. R., *J. Biol. Chem.*, **237**, 1787 (1962)
42. Vagelos, P. R., Alberts, A. W., and Martin, D. B., *Biochem. Biophys. Res. Commun.*, **8**, 4 (1962)
43. Lynen, F., Matsuhashi, M., Numa, S., and Schweizer, E. In *The Control of Lipid Metabolism* (Biochemical Society Symposium, No. 24), p. 43. Ed. by Grant, J. K., Academic Press Inc., London and New York, 1963
44. Vagelos, P. R., Alberts, A. W., and Martin, D. B., *J. Biol. Chem.*, **238**, 533 (1963)
45. Waite, M., and Wakil, S. J., *J. Biol. Chem.*, **237**, 2750 (1962)
46. Gregolin, C., Ryder, E., and Lane, M. D., *J. Biol. Chem.*, **243**, 4227 (1968)
47. Gregolin, C., Ryder, E., Warner, R. C., Kleinschmidt, A. K., Chang, H. C., and Lane, M. D., *J. Biol. Chem.*, **243**, 4236 (1968)
48. Gregolin, C., Ryder, E., Kleinschmidt, A. K., Warner, R. C., and Lane, M. D., *Proc. Nat. Acad. Sci. U.S.*, **56**, 148, 1751 (1966)
49. Numa, S., Ringelmann, E., and Reidel, B., *Biochem. Biophys. Res. Commun.*, **24**, 750 (1966)
50. Goto, T., Ringelmann, E., Reidel, B., and Numa, S., *Life Sci.*, **6**, 785 (1967)
51. Numa, S., Ringelmann, E., and Lynen, F., *Biochem. Z.*, **343**, 243, 258 (1965)
52. Numa, S., Bortz, W. M., and Lynen, F., *Advan. Enz. Reg.*, **3**, 407 (1965)
53. Greenspan, M. D., and Lowenstein, J. M., *J. Biol. Chem.*, **243**, 6273 (1968)
54. Miller, A. L., and Levy, H. R., *J. Biol. Chem.*, **244**, 2334 (1969)
55. Spencer, A. F., and Lowenstein, J. M., *Biochem. J.*, **103**, 342 (1967)
56. Denton, R. M., Yorke, R. E., and Randle, P. J., *Biochem. J.*, **100**, 407 (1966)
57. Denton, R. M., and Halperin, M. L., *Biochem. J.*, **110**, 27 (1968)
58. Margolis, S. A., and Baum, H., *Arch. Biochem. Biophys.*, **114**, 445 (1966)
59. Smith, S., Easter, D. J., and Dils, R., *Biochim. Biophys. Acta*, **125**, 445 (1966)
60. Den. H., and Klein, H. P., *Biochim. Biophys. Acta*, **49**, 429 (1961)
61. Iliffe, J., and Myant, N. B., *Biochem. J.*, **117**, 385 (1970)
62. Foster, D. W., and McWhorter, W. P., *J. Biol. Chem.*, **244**, 260 (1969)
63. Pirson, W., and Lynen, F., *Hoppe-Seyler's Z. Physiol. Chem.*, **352**, 797 (1971)
64. Denton, R. M., and Randle, P. J., *Biochem. J.*, **100**, 420 (1966)
65. Majerus, P. W., and Kilburn, E., *J. Biol. Chem.*, **244**, 6254 (1969)
66. Nakanishi, S., and Numa, S., *Eur. J. Biochem.*, **16**, 161 (1970)
67. Nakanishi, S., and Numa, S., *Proc. Nat. Acad. Sci. U.S.*, **68**, 2288 (1971)
68. Butterworth, P. H. W., Yang, P. C., Bock, R. M., and Porter, J. W., *J. Biol. Chem.* **242**, 3508 (1967)
69. Plate, C. A., Joshi, V. C., Sedgwick, B., and Wakil, S. J., *J. Biol. Chem.*, **243**, 5439 (1968)
70. Lust, G., and Lynen, F., *Eur. J. Biochem.*, **7**, 68 (1968)
71. Bressler, R., and Wakil, S. J., *J. Biol. Chem.*, **237**, 1441 (1962)
72. Nixon, J. E., Putz, G. R., and Porter, J. W., *J. Biol. Chem.*, **243**, 5471 (1968)
73. Yalpani, M., Willecke, K., and Lynen, F., *Eur. J. Biochem.*, **8**, 495 (1969)
74. Brock, D. J., and Bloch, K., *Biochem. Biophys. Res. Commun.*, **25**, 473 (1966)
75. Katz, J., Landau, B. R., and Bartsch, G. E., *J. Biol. Chem.*, **241**, 727 (1966)
76. Katz, J., and Rognstad, R., *J. Biol. Chem.*, **241**, 3600 (1966)
77. Shrago, E., Lardy, H. A., Nordlie, R. C., and Foster, D. O., *J. Biol. Chem.*, **238**, 3188 (1963)
78. Weber, G., and Convery, H. J. H., *Life Sci.*, **5**, 1139 (1966)

79. Matthes, K. J., Abraham, S., and Chaikoff, I. L., *Biochim. Biophys. Acta*, **71**, 568 (1963)
80. Lynen, F., *Fed. Proc.*, **20**, 941 (1961)
81. Schweizer, E., Lerch, I., Kroeplin-Rueff, L., and Lynen, F., *Eur. J. Biochem.*, **15**, 472 (1970)
82. Lynen, F., *Biochem. J.*, **102**, 381 (1967)
83. Sumper, M., Oesterhelt, D., Riepertinger, C., and Lynen, F., *Eur. J. Biochem.*, **10**, 377 (1969)
84. Ayling, J., Pirson, R., and Lynen, F., *Biochemistry*, **11**, 526 (1972)
85. Smith, S., and Dils, R., *Biochim. Biophys. Acta*, **116**, 23 (1966)
86. Dils, R., and Popják, G., *Biochem. J.*, **83**, 41 (1962)
87. Becker, M. E., and Kumar, S., *Biochemistry*, **4**, 1839 (1965)
88. Carey, E. M., Dils, R., and Hansen, H. J. M., *Biochem. J.*, **117**, 633 (1970)
89. Bishop, C., Davies, T., Glascock, R. F., and Welch, V. A., *Biochem. J.*, **113**, 629 (1969)
90. Phillips, G. T., Nixon, J. E., Dorsey, J. A., Butterworth, P. H. W., Chesterton, C. J., and Porter, J. W., *Arch. Biochem. Biophys.*, **138**, 380 (1971)
91. Barnes, E. M., and Wakil, S. J., *J. Biol. Chem.*, **243**, 2955 (1968)
92. Chang, H., Seidman, I., Teebor, G., and Lane, M. D., *Biochem. Biophys. Res. Commun.*, **28**, 682 (1967)
93. Del Boca, J., and Flatt, J. P., *Eur. J. Biochem.*, **11**, 127 (1969)
94. Crofford, O. B., and Renold, A. E., *J. Biol. Chem.*, **240**, 3237 (1965)
95. Flatt, J. P., *J. Lipid Res.*, **11**, 131 (1970)
96. Schweizer, E., Piccinini, F., Duba, C., Günther, S., Ritter, E., and Lynen, F., *Eur. J. Biochem.*, **15**, 483 (1970)
97. Burton, D. N., Collins, J. M., and Porter, J. W., *J. Biol. Chem.*, **244**, 1076 (1969)

CHAPTER 5

Biosynthesis of Triglycerides and Complex Lipids

Fatty acids are not synthesized for use *per se* but rather for conversion into various classes of lipid, a term used to describe a wide-ranging group of natural compounds. These all have their especial role to play within the cell whether it is structural or metabolic. It is therefore felt that this text would not be complete without some attempt to discuss their biosynthesis but as this topic has developed into a rapidly expanding field, emphasis will tend to be given to general concepts and to some recent progress in the mode of formation of triglycerides, phospholipids and sphingolipids. A short account of the function and localization of the main classes of lipid, however, will be given first.

Triglycerides (1,2,3-triacylglycerol, (I)) represent the major food reserve in animals and are commonly known as fats or oils, depending on their physical state at room temperature. This in turn rests upon the degree of unsaturation of the component fatty acids which are present in ester linkage with the three hydroxyl groups of glycerol. Although they are found in all tissues, triglycerides are deposited predominantly in localized areas of adipose tissue distributed throughout the body, for storage purposes. Their activity in this respect resides in the energy released during oxidation of the aliphatic chain within the constituent fatty acids. Triglycerides are primarily called upon to act in this capacity during conditions of starvation and other situations in which glucose metabolism is affected or where a calorific deficit is set up as occurs during exercise. In these circumstances, fatty acids are needed to satisfy energy requirements and are formed by hydrolysis. They are then released into the plasma where they bind to albumin and are transported to certain tissues including liver and muscle that can utilize them as an energy source.

Structurally many phospholipids (e.g. (II)) are derived from diglyceride (diacylglycerol), phosphate and a suitable base such as choline. The related sphingolipids will be dealt with separately and contain a sphingenine (III) residue instead of the glycerol backbone. The phosphate group is attached to the alcohol group of the polar nitrogen-containing moiety (or inositol or glycerol) through an ester bridge. The products may therefore be acidic (due to the presence of phosphate anions) or neutral at physiological pH values depending on the nature of this unit. In the formulae, R.CO represents

$$^1CH_2.O.CO.R_1$$
$$R_2.CO.O.^2CH$$
$$^3CH_2.O.CO.R_3$$

$$^1CH_2.O.CO.R_1$$
$$R_2.CO.O.^2CH \quad O$$
$$^3CH_2.O.\overset{\parallel}{P}.O.CH_2.CH_2.N^+(CH_3)_3$$
$$O^-$$

(I)   (II)

long-chain acyl residues; R_1 and R_3 tend to be saturated groups while R_2 tends to be unsaturated. Phospholipids possess at different points in the molecule ionic and hydrophobic (due to the long-chain R groups) portions which confer certain surface-active properties on them and render them particularly suitable for their physiological role. They occur naturally within the cell bound to proteins but are readily obtained after extraction with organic solvents. Phospholipids have attracted great attention in recent years for two principal reasons: significant improvements in the techniques available for their isolation and fractionation from each other, and an appreciation that phospholipid–protein complexes may be considered as the major building blocks of many cell membranes. These lipoprotein complexes are found universally in membranes and possess important structural functions. A model for biological membranes was proposed many years ago by Danielli and Davson[1] in which two monolayers of phospholipid were placed with their polar ends directed towards globular protein (and held by interaction with charged groups), whereas the hydrophobic long-chain acyl groups were associated with each other by means of van der Waals' forces. A modification of this structure for myelin was later proposed[2,3] in which it was suggested that a layer of neutral lipid (cholesterol) lay between the non-polar portions of the phospholipids. However, this model is inadequate to explain certain membrane properties and different views giving somewhat modified structures, for membranes other than that found in myelin, have since been expressed. These have now received wider support and have been collated in a recent review article.[4]

The major biological function of phospholipids is to act as structural components of the membranes of cells and subcellular organelles. They form as lipoprotein complexes the main constituents of mitochondria and endoplasmic reticulum. They are likewise essential for the integration and stability of such enzymic sequences in the mitochondria as the electron transport system and oxidative phosphorylation and, in addition, for transport of ions and other metabolites. Lecithin (phosphatidyl choline, (II)) is the main phospholipid distributed in animal tissues. Phospholipids also occur in high concentration in brain and the nervous system, muscle and the seeds of higher plants with phosphatidyl ethanolamine as the principal component. Much of this lipid, however, is found in a modified form as plasmalogen (p. 96).

The sphingolipids are especially abundant in brain and nervous tissue and comprise the major component of myelin. Chemically, they are characterized by the presence of a C_{18} amino alcohol, sphing-4-enine[5] (sphingosine; trans-D-erythro-1,3-dihydroxy-2-amino octadec-4-ene; (III)), to which a long-chain acyl residue is bound in amide linkage. The primary alcohol

$$^{18}CH_3.[CH_2]_{12}.\overset{t}{CH}:\,^4CH.\,^3CH(OH).\,^2CH(NH_2).\,^1CH_2OH$$

(III)

group at C-1 may be attached by ester linkage to phosphorylcholine (sphingomyelin, (IV)), a product that may therefore also be classified as a phospholipid, or by a glycosidic bond to β-galactose or β-glucose (cerebroside, (V)). The related alcohol, 4-hydroxysphinganine (phytosphingosine; 4-hydroxy-dihydrosphingosine) is a constituent of plant glycosphingolipids.

$$CH_3.[CH_2]_{12}.CH:CH.CH(OH).CH(\overset{|}{NH}).CH_2.O.\overset{O}{\overset{||}{P}}.O.CH_2.CH_2.N^+.(CH_3)_3$$
$$R.CO \qquad O^-$$

(IV)

$$CH_3.[CH_2]_{12}.CH:CH.CH(OH).CH(\overset{|}{NH}).CH_2.O-\beta\text{-galactose }(\beta\text{-glucose})$$
$$R.CO$$

(V)

BIOSYNTHESIS OF TRIGLYCERIDES (TRIACYLGLYCEROL)

The principal features of the pathways responsible for the synthesis of triglycerides and phospholipids were established some time ago, largely as a result of the outstanding contributions by Kennedy and his colleagues. The early work gave rise to the realization that L-α-phosphatidate (VI) and 1,2-diglyceride occupied intermediary positions in the pathways leading to the synthesis of many classes of lipid.[6] A recent commission on nomenclature[5] has advised that the structure of glycerol derivatives may be unambiguously expressed by a system of stereospecific numbering. If the secondary alcohol group attached at C-2 appears on the left in the normal Fischer projection (that is, in the L-configuration), the carbon atom above this (bearing one of the primary hydroxyl groups) is designated C-1 in this nomenclature. On this basis, the two metabolites mentioned above would be named 1,2-diacyl-sn-glycerol 3-phosphate (VI) and 1,2-diacyl-sn-glycerol. This system will be used when there might otherwise be possibility of confusion.

$$^1CH_2.O.CO.R_1$$
$$R_2.CO.O.^2CH$$
$$^3CH_2.O.\overset{O}{\overset{||}{P}}.OH$$
$$O^-$$
(VI)

The major sites of synthesis of triglycerides in animal tissues occur in the liver (from where they are transported in the plasma as very low density lipoproteins), adipose tissue (and intestines) by a process involving esterification of sn-glycerol 3-phosphate, with phosphatidate (VI) as intermediate. In the intestines, however, another mechanism in which monoglycerides are acylated directly operates as the main pathway.

Liver and Adipose Tissue (and Microorganisms)

Early studies[7,8] on the biosynthesis of triglycerides in rat and chicken liver preparations demonstrated that they were not formed by a simple reversal of lipolysis, achieved by hydrolytic cleavage of acyl glycerol bonds. ATP was essential for the formation of glycerol 3-phosphate and the activation of long-chain fatty acids by acyl-CoA synthetase [fatty acid:CoA ligase (AMP), EC 6.2.1.3]. It was readily established that the acyl transferases involved were particulate, and that the microsomal fraction proved more effective than mitochondria in promoting synthesis.[9] Kennedy and co-workers[8,10] proposed the following series of reactions as the essential steps in the synthesis of triglycerides (equations (1)–(4)):

$$\text{Glycerol} + \text{ATP} \longrightarrow \text{glycerol 3-phosphate} + \text{ADP} \tag{1}$$

$$\text{Glycerol 3-phosphate} + \text{acyl.S.CoA} \rightleftharpoons$$
$$\text{CoASH} + \text{monoacylglycerol phosphate} \xrightleftharpoons{\text{acyl.S.CoA}} \text{phosphatidate} \tag{2}$$

$$\text{Phosphatidate} + \text{H}_2\text{O} \longrightarrow \text{1,2-diglyceride} + \text{phosphate} \tag{3}$$

$$\text{1,2-Diglyceride} + \text{acyl.S.CoA} \rightleftharpoons \text{triglyceride} + \text{CoASH} \tag{4}$$

The formation of glycerol phosphate is an obligatory step in the synthesis of triglyceride. It may be formed by phosphorylation of glycerol (derived from hydrolytic breakdown of lipids) with ATP in a reaction controlled by glycerol kinase (ATP:glycerol phosphotransferase, EC 2.7.1.30) (equation (1)) but more usually by reduction of dihydroxyacetone phosphate, generated by the glycolytic sequence of reactions, with the NAD-linked dehydrogenase (L-(sn)-glycerol 3-phosphate:NAD oxidoreductase, EC 1.1.1.8) (equation (5)):

$$\text{Dihydroxyacetone phosphate} + \text{NADH} + \text{H}^+ \rightleftharpoons \text{glycerol 3-phosphate} + \text{NAD}^+ \tag{5}$$

Glycerol kinase has limited distribution in animal tissues but is found in liver,[11,12] kidney[12] and intestinal mucosa,[13] mainly in the cell sap. It is essentially absent from adipose tissue[14] and glycerol phosphate must therefore be formed by the reaction described in equation (5). It subsequently reacts with two molecules of acyl-CoA to generate phosphatidate, a key intermediate in the overall process (equation (2)); the enzyme responsible is acyl-CoA:glycerol 3-phosphate O-acyltransferase (EC 2.3.1.15). Guinea-pig liver microsomes were used to study this two-stage reaction with [¹⁴C]-palmitoyl-CoA and either glycerol phosphate or 1-acylglycerol phosphate (lysophosphatidate) as acceptor substrates.[15] These reactions were catalysed by distinct acyl transferases since the first step only was dependent on an active thiol centre and was inhibited by thiol group reagents; acylation of lysophosphatidate remained unaffected. More recently, lysophosphatidate was identified as the product from rat liver microsomes under conditions in which monoacyl transferase activity was prevented.[16] Palmitoyl and oleoyl groups were preferentially attached to C-1 and C-2 of sn-glycerol 3-phosphate respectively. However, an enzyme from rat brain possesses the capacity to acylate both hydroxyl groups.[17] The L-configuration present in glycerol phosphate is retained throughout these and subsequent reactions.

Similar investigations with microorganisms confirmed that the enzymes involved in phosphatidate synthesis were particulate but indicated that two enzymes were involved. Particulate preparations derived from E. coli membranes could transfer the acyl residue from palmitoyl-ACP and palmitoyl-CoA to glycerol 3-phosphate during synthesis of lysophosphatidate and subsequent formation of phosphatidate (further acylation) or monoglyceride (dephosphorylation).[18] Palmitoyl-ACP gave rise mainly to lysophosphatidate but phosphatidate was formed in the presence of unsaturated acyl-ACP substrates. A similar monoacyl transferase preparation from C. butyricum converted palmitoyl-CoA and glycerol phosphate into lysophosphatidate but synthesis was stimulated on addition of ACP (obtained from E. coli), suggesting that the true substrate was the ACP thioester.[19]

Palmitoyl-CoA acyltransferase (EC 2.3.1.15) activity was also demonstrated in yeast microsomes.[20] The enzyme was sensitive to thiol group reagents (N-ethylmaleimide and iodoacetamide) but preincubation with palmitoyl-CoA or oleoyl-CoA strongly protected the enzyme. The product was tentatively identified after thin-layer chromatography as a monoacylated glycerophosphate (lysophosphatidate) and a thiol-bound derivative was proposed as an intermediate (equations (6) and (7)):

$$\text{Palmitoyl.S.CoA} + \text{HS.enzyme} \rightleftharpoons \text{palmitoyl.S.enzyme} + \text{CoASH} \qquad (6)$$

$$\text{Palmitoyl.S.enzyme} + \text{glycerol phosphate} \rightleftharpoons \text{lysophosphatidate} + \text{HS.enzyme} \qquad (7)$$

Many in vitro systems are active for the synthesis of phosphatidate but tissue levels are invariably low. This intermediate has a high turnover rate and is rapidly metabolized in vivo for lipid synthesis. It is cleaved by a specific phosphatase (phosphatidate phosphohydrolase, EC 3.1.3.4) to generate 1,2-diglyceride and inorganic phosphate[21] (equation (3)). The diglyceride is finally esterified with a third molecule of acyl-CoA to give triglyceride[8] by means of a diglyceride acyltransferase (acyl-CoA:1,2-diglyceride O-acyltransferase, EC 2.3.1.20) (equation (4)).

The distinctive pattern of triglycerides and phospholipids in tissues indicates that glycerol phosphate is preferentially reacted with saturated acyl groups at C-1 and unsaturated moieties at the C-2 ester position.[22,23]

This information was obtained after identification of the acids released by pancreatic lipase and phospholipase treatment[24] and confirmed following *in vitro* studies on the formation of phosphatidate and phospholipids.[25] Thus phosphatidate itself possesses asymmetry with regard to the status of the component fatty acids. Moreover, some plant triglycerides contain almost exclusively unsaturated acyl groups at C-2.[23,26] These results confirm that two enzymes with different positional specificities are generally involved in the formation of phosphatidate and hence diglycerides, triglycerides and phospholipids. Several diglycerides were tested for their efficiency as acceptors and 1,2-diolein (1,2-dioleoyl-*sn*-glycerol) proved the most effective in this respect,[8] but the activity of individual diglycerides in these experiments might have been influenced to some extent by solubility and dispersion factors. Unsaturated acyl residues confer greater solubility and hence emulsifying properties on their derived lipids. Moreover, it was noted that their conversion into lecithin (in the presence of CDP-choline) was considerably lower. Examination has since revealed[27] that mammalian phospholipids possess a greater degree of unsaturation in the fatty acids attached to C-2 than related triglycerides, implying that the enzymes concerned with triglyceride synthesis may have different specificities with regard to chain-length and degree of unsaturation of the substrate fatty acids. However, this selectivity is probably related to the participation of a mono-acyl-diacyl phosphatide cycle[15,28,29] which will be mentioned later (p. 95).

CDP-derivatives are directly involved in phospholipid synthesis but evidence presented by Marinetti's group[30,31] implicated CTP as an effector substance for the formation of neutral lipids. CTP (and CMP) strongly stimulated synthesis of diglycerides and triglycerides from [^{14}C]glycerol in rat liver homogenates at the expense of lecithin, a property specific for cytosine nucleotides. The average of a number of similar experimental results illustrating this effect is given in Table 5.1.

Studies with homogenates of rat adipose tissue indicated that palmitate was incorporated into triglyceride in the presence of glycerol phosphate and suitable cofactors.[32] The requirement for glycerol phosphate could be

Table 5.1   Effect of CTP on the incorporation of [^{14}C]glycerol into diglycerides, triglycerides and lecithin (from Marinetti, 1970)[31]

	Radioactivity		
	Diglyceride	Triglyceride (counts/min)	Lecithin
Control	4,030	3,450	21,800
Plus CTP	10,320	12,950	5,200

met with various glycolytic intermediates plus NADH but glycerol itself was ineffective as a precursor due to the absence of glycerol kinase.[14] Several groups of investigators[33,33a] later confirmed that the overall process of esterification was associated with microsomal preparations but one report[33a] did indicate that mitochondria were considerably more effective with free fatty acids as substrate. The acylating enzymes in chicken adipose tissue were also located in the particulate fraction and the diglyceride acyltransferase acted preferentially on substrates containing at least one unsaturated fatty acid.[34]

Intestines

It was first established in Hübscher's Laboratory[35] and later confirmed by others[36,37] that triglycerides are synthesized in the intestines mainly (approximately 80–100 per cent) *via* a pathway involving direct acylation of 2-monoglycerides available from dietary sources with suitable acyl-CoA substrates. Phosphatidate does not act as an intermediate in this sequence. These observations were made during studies with intestinal epithelial cells from rat and rabbit mucosa. The microsomal fraction was again particularly effective in terms of specific activity and total capacity. Since diglycerides behave as precursors in this process in addition to that involving glycerol phosphate, an evaluation of possible equilibration of the products formed by the two pathways was made. The acylating enzymes in intestinal cells are relatively non-specific and do not favour transfer of unsaturated acyl groups to the C-2 hydroxyl position, with the result that triglycerides synthesized by this pathway tend to be atypical. The characteristic pattern of fatty acids is controlled by the specificity of these enzymes and this is retained in the intestinal triglycerides. Thus the two pools apparently remain independent.[38]

Rao and Johnston[39] purified the enzymes concerned in the monoglyceride pathway [acyl-CoA synthetase (EC 6.2.1.3) and the monoglyceride- and diglyceride acyltransferases] from mucosal microsomes and noted that they purified simultaneously. The functional moiety of coenzyme A also remained enzyme-bound during the entire reaction sequence. These facts indicated that the enzymes might behave as a multienzyme complex to which the substrates and intermediate product were bound. Thus they argued that an ACP-like molecule with its acyl derivative might be implicated in this process. It had previously been shown[40] that extracts of *E. coli* utilized palmitoyl-ACP as an acyl substrate in the formation of phosphatidate.

Control Aspects

A rapid rate of triglyceride synthesis is obviously dependent on an adequate supply of glycerol phosphate and acyl-CoA substrates. These in turn are ultimately derived from excess carbohydrate in the diet *via* the formation of dihydroxyacetone phosphate and acetyl-CoA. In addition, triglycerides in adipose tissue and liver, for instance, do not remain static and are constantly mobilized and resynthesized as the nutritional situation demands.

Synthesis is also profoundly affected by hormonal status. Acyltransferase (EC 2.3.1.15) level increases greatly in mammary gland tissue at parturition.[41] Insulin exerts a number of short- and long-term effects by increasing the permeability of certain tissues to glucose and hence the supply of precursors by means of an increased flux through glycolysis. This factor is especially important in adipose tissue. However, the overall process of conversion of glucose to fatty acid or triglyceride generates ATP and is therefore limited by the energy requirements of this tissue (and hence availability of ADP after recycling).[42] This problem has been briefly touched upon at the end of the previous chapter. Insulin also acts at the genetic level[43] by stimulation of de novo synthesis of ATP citrate lyase (EC 4.1.3.8), acetyl-CoA carboxylase (EC 6.4.1.2), hexose monophosphate dehydrogenases (EC 1.1.1.49 and 1.1.1.44) and malate dehydrogenase (decarboxylating) (EC 1.1.1.40). More-over, it also inhibits adenyl cyclase activity. Consequently the activation of adipose tissue (mobilizing) lipase (EC 3.1.1.3) by cyclic-3',5'-AMP becomes limiting and controls the rate at which triglycerides are hydrolysed, their products mobilized by release into the bloodstream and transported to the liver or other tissues which can oxidize them. Similarly the lipoprotein lipase (clearing factor lipase) associated with the endothelial cells in the capillaries surrounding adipose tissue is affected by cyclic-AMP in a comple-mentary manner (inhibition).[44] Accordingly, this enzyme is activated in the presence of insulin (which means in effect after a meal) when the formation of cyclic-AMP is discouraged. The liberated fatty acids may then be transported into the adipose tissue cells. Insulin and situations in which normal dietary conditions prevail (the 'fed state') therefore tend to stimulate net synthesis of triglyceride. In contrast adrenaline, glucagon, ACTH and adrenal corti-coids activate the cyclase and thereby encourage (mobilizing) lipase activity.[31] A similar situation occurs in diabetes or the fasted state where lack of insulin increases lipolysis; insulin is the only hormone that favours storage. More-over, reduction of de novo lipogenesis in adipose tissue during fasting is related to a decrease in the formation of acetyl-CoA[45] and to the ability of free fatty acids to act as substrates for the production of NADH and ATP.[42]

It has already been mentioned[30,31] that CTP and CMP markedly affect the relative rates of neutral lipid and lecithin synthesis in the liver (Table 5.1). Since these cytosine nucleotides act as precursor and end-product respec-tively for phospholipid biosynthesis, a rationale for a possible physiological regulation of lipid formation on a complementary basis may be readily envisaged.

BIOSYNTHESIS OF PHOSPHOLIPIDS

The term phospholipid includes a large variety of lipids all of which are highly polar because of the ionic phosphate residue they possess. Biosynthetically they may be divided into a number of classes dependent on whether they derive from cytidinediphospho(CDP)-linked bases plus 1,2-diacyl-sn-glycerol, the 1-alkyl (or 1-alkenyl)-2-acyl-sn-glycerol ether analogue or from CDP-diglyceride (1,2-diacyl-sn-glycerol 3-pyrophosphorylcytidine). The various modes of synthesis that lead to the formation of these phospholipids will now be discussed.

Biosynthetic Pathways Involving Cytidinediphospho(CDP)-Linked Bases

Kornberg and Pricer[46] first demonstrated that phosphorylcholine, doubly labelled with ^{14}C and ^{3}H, was converted into a lipid [later identified as lecithin (II)] by a liver enzyme with retention of the original ^{14}C:^{3}H ratio. Thus the choline portion at least of this substrate was incorporated as a unit. The foundation for the discovery of the pathways involved in phospholipid synthesis was firmly laid when CTP was identified as the nucleotide responsible for the formation of acceptor molecules.[47] It is interesting to record that this breakthrough was established after noting that impure samples of ATP (that contained trace amounts of CTP as phosphoryl donor) had proved considerably more effective in promoting lecithin synthesis than either the crystallized preparation or a generating system for ATP. Kennedy and Weiss[47] then isolated cytidinediphosphocholine (CDP-choline) and the ethanolamine derivative from tissues such as liver and yeast and, in addition, prepared it enzymically from CTP and the phosphorylated bases. The products could be converted into phospholipid in the presence of 1,2-diglyceride. Experiments with rat brain homogenates confirmed[48] that ^{14}C-labelled glycerol phosphate, phosphorylcholine and CDP-choline, but not glycerol [^{32}P]phosphate, were incorporated into lecithin. Phospholipid synthesis therefore diverges from that of neutral triglycerides at some stage. CDP-linked bases also participate in the formation of the glyceryl ether analogues and related plasmalogens.

The initial reaction in the biosynthesis of lecithin involves the phosphorylation of choline by an ATP-mediated kinase (ATP:choline phosphotransferase, EC 2.7.1.32). The product, phosphorylcholine (VII), then reacts with CTP under the influence of CTP:cholinephosphate cytidylyl-transferase (EC 2.7.7.15) to give CDP-choline (VIII) with the elimination of pyrophosphate.[49] This nucleotide derivative finally reacts with a diglyceride acceptor in a transfer catalysed by CDP-choline:1,2-diglyceride cholinephosphotransferase (EC 2.7.8.2) to give lecithin (II) and CMP.[50] The steps leading to the synthesis of lecithin are depicted in Scheme 5.1 and indicate how the terminal phosphate of ATP is retained in the products (VII), (VIII)

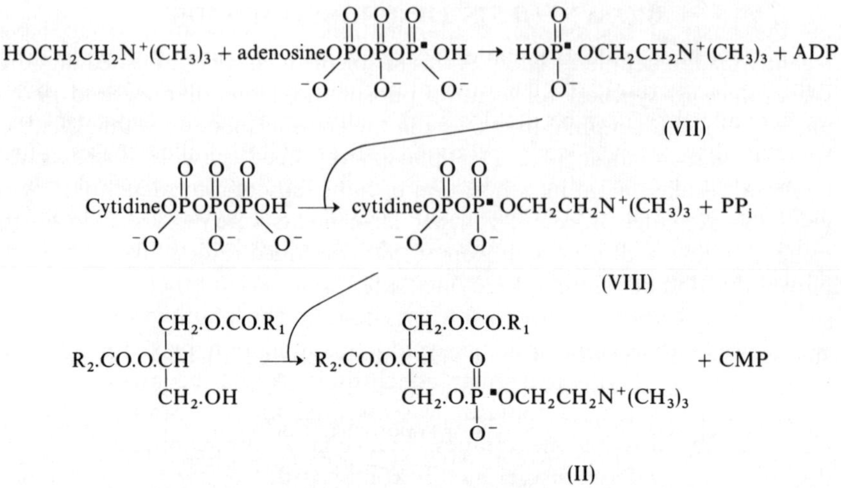

$$HOCH_2CH_2N^+(CH_3)_3 + adenosineO\overset{O}{\underset{\underset{O^-}{|}}{\overset{||}{P}}}O\overset{O}{\underset{\underset{O^-}{|}}{\overset{||}{P}}}O\overset{O}{\underset{\underset{O^-}{|}}{\overset{||}{P}}}{}^\bullet OH \rightarrow HO\overset{O}{\underset{\underset{O^-}{|}}{\overset{||}{P}}}{}^\bullet OCH_2CH_2N^+(CH_3)_3 + ADP$$

(VII)

$$CytidineO\overset{O}{\underset{\underset{O^-}{|}}{\overset{||}{P}}}O\overset{O}{\underset{\underset{O^-}{|}}{\overset{||}{P}}}O\overset{O}{\underset{\underset{O^-}{|}}{\overset{||}{P}}}OH \longrightarrow cytidineO\overset{O}{\underset{\underset{O^-}{|}}{\overset{||}{P}}}O\overset{O}{\underset{\underset{O^-}{|}}{\overset{||}{P}}}{}^\bullet OCH_2CH_2N^+(CH_3)_3 + PP_i$$

(VIII)

$$\begin{array}{l}CH_2.O.CO.R_1 \\ R_2.CO.O.\underset{|}{C}H \\ CH_2.OH\end{array} \longrightarrow \begin{array}{l}CH_2.O.CO.R_1 \\ R_2.CO.O.\underset{|}{C}H \quad O \\ CH_2.O.\overset{||}{\underset{\underset{O^-}{|}}{P}}{}^\bullet OCH_2CH_2N^+(CH_3)_3\end{array} \quad + CMP$$

(II)

Scheme 5.1   Involvement of nucleotides in the biosynthesis of lecithin(II). PP_i denotes $HP_2O_7^{3-}$ (pyrophosphate)

and (II). Formation of the related phosphatidylethanolamine proceeds in a similar manner by donation of the phosphorylethanolamine portion to 1,2-diglyceride.[51] Cytidylyl transferase (EC 2.7.7.14) and ethanolamine phosphotransferase (EC 2.7.8.1) enzymes are responsible[52] but it has not been settled whether these differ from those utilizing choline.

Phosphatidylethanolamine may also be formed in the liver through decarboxylation of phosphatidylserine,[53] itself derived in this tissue by an exchange reaction between phosphatidylethanolamine and L-serine. The net effect of these reactions is the constitution of a phospholipid cycle that results in the decarboxylation of serine. This phospholipid in turn may be converted directly into lecithin in microsomal preparations[51,54] by sequential transmethylation of the amino group with S-adenosylmethionine as methyl donor (Scheme 5.2).

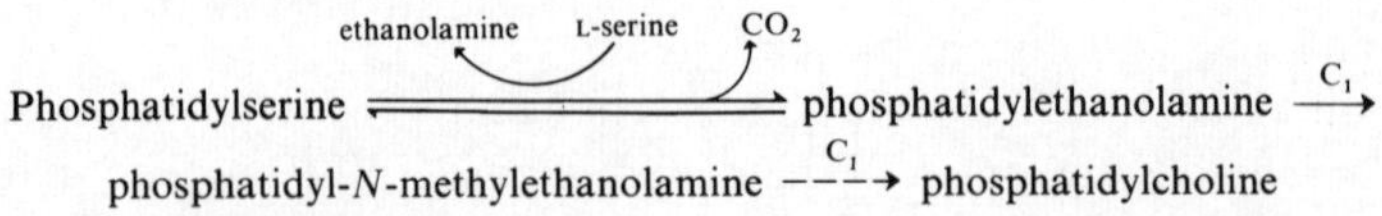

$$Phosphatidylserine \underset{ethanolamine}{\overset{L\text{-serine} \quad CO_2}{\rightleftharpoons}} phosphatidylethanolamine \xrightarrow{C_1}$$

$$phosphatidyl\text{-}N\text{-methylethanolamine} \dashrightarrow{}^{C_1} phosphatidylcholine$$

Scheme 5.2   Interrelationships between phosphatidyl bases in the liver

The enzymes responsible for hepatic synthesis of CDP-choline (EC 2.7.7.15) and its subsequent reaction with diglyceride (EC 2.7.8.2) are localized in the microsomes.[55] Data presented by many workers in early studies suggested that mitochondria also possessed these enzymes but the limited synthesis that was apparent in these organelles was later explained on the basis of microsomal contamination.[56,57] Dawson's group have recently made a detailed study[58] of this problem and have established that the site of *de novo* synthesis of lecithin, phosphatidylethanolamine and phosphatidylinositol in animal liver exists in the endoplasmic reticulum. Incubation of mitochondria plus [^{32}P]phosphate together with microsomes, however, resulted in an incorporation of radioactivity into these lipids by a reversible exchange process.[56] Further experiments were performed in which isolated mitochondria or microsomes that contained radioactive lecithin or phosphatidylethanolamine were incubated with each other. The results showed that transfer of the freshly synthesized phospholipid between the two subcellular fractions had occurred.[57] Examination by pulse-labelling techniques[58] gave rise to similar conclusions. A specific carrier protein present in the soluble cytoplasm was essential for the transfer from the synthetic site across the membranes.[59] Once in the mitochondria, however, the lipids may undergo further modification possibly by means of exchange, acylation of lysolecithin or *N*-methylation reactions. Thus, lecithin, a major phospholipid in liver mitochondria, appears to be synthesized outside this organelle, a fact that poses interesting questions concerning the biogenesis of mitochondria.

The precise nature of the fatty acid groups (and bases) present in the phospholipid from a given membrane seems to reflect a specific requirement for the tissue and subcellular organelle from which it has been derived. Species of a structurally distinct type of phospholipid may vary according to the chain-length, degree of unsaturation and distribution of the fatty acids attached to the glycerol 'backbone'. This metabolic heterogeneity is responsible for the alteration in physicochemical characteristics and hence the properties of the membranes in which these lipids are found and arises in part from the relative specificity of the acyl-transferases (equation (2)) utilized for their synthesis. An important factor, however, relates to the selective specificity of enzymes (e.g. acyl-CoA : lysolecithin acyltransferase) that act on 1-acyl- and 2-acylglycerol 3-phosphorylcholine and their ethanolamine analogues.[15] These monoacyl derivatives are preferentially esterified with unsaturated[60] (especially C_{20} polyunsaturated acids)[29] and saturated acids respectively. They are formed after deacylation by acyl hydrolases (phospholipases)[22] and the combined process of hydrolysis and re-esterification constitutes a monoacyl-diacyl phosphatide cycle.

De novo synthesis of phosphatidate,[61] diphosphoinositide[62,63] and sphingomyelin[64] (IV) may occur in both mitochondria and microsomes. Phosphatidate formation is localized in the outer mitochondrial membrane, as assessed by electron microscopy and comparison with marker enzymes. In one particular study, great care was exercised to ensure that valid criteria were used to differentiate between the outer membrane and smooth and rough endoplasmic reticulum.[61]

Ether-Linked Lipids

Phospholipids containing alkyl residues bound by ether linkage to the C-1 hydroxyl group of glycerol instead of the more usual ester bond are widespread and quantitatively of considerable importance. They may comprise 20 per cent of the total phospholipid[65] and possess either a saturated alkyl group (glyceryl ether (IX)) or a cis-α-β-unsaturated (vinylic) derivative (plasmalogen; e.g. 1-alk-1'-enyl-2-acyl-sn-glycerol 3-phosphorylcholine, (X)).

$$^1CH_2.O.CH_2.CH_2.R_1$$
$$R_2.CO.O.^2CH$$
$$\begin{matrix} & O \\ & \| \end{matrix}$$
$$^3CH_2.O.\overset{\|}{P}.O\text{—choline}$$
$$\overset{|}{O^-} \quad \text{(ethanolamine)}$$
$$\text{(IX)}$$

$$^1CH_2.O.CH:CH.R_1$$
$$R_2.CO.O.^2CH$$
$$\begin{matrix} & O \\ & \| \end{matrix}$$
$$^3CH_2.O.\overset{\|}{P}.O\text{—choline}$$
$$\overset{|}{O^-} \quad \text{(ethanolamine)}$$
$$\text{(X)}$$

CDP-choline or CDP-ethanolamine also react with a preformed 1-O-alkenyl-2-acylglycerol as acceptor to give a plasmalogen.[66] These lipids are therefore related structurally and biosynthetically to lecithin and phosphatidylethanolamine. It was later confirmed that [^{14}C]palmitate and stearate (and stearaldehyde) were readily incorporated by the terrestrial slug[67] and starfish[68] into ether-linked phospholipids, principally the glyceryl ether components. The unsaturated plasmalogens consistently had lower specific radioactivities, suggesting that they might possibly be formed by desaturation of (IX). Thompson[69] conducted experiments in which doubly-labelled chimyl alcohol (1-O-[^{14}C]hexadec-2-[^{3}H]glycerol) was fed to slugs. He and others[70] observed that radioactivity from ^{14}C was effectively incorporated into the glyceryl ether and plasmalogen lipids. Moreover, the ^{14}C/^{3}H ratio present in the substrate was retained in both lipids (but with some dilution of ^{3}H in plasmalogen), confirming a precursor-product relationship.[70] Stereospecifically labelled substrates ([^{14}C, 1-^{3}H]- and [^{14}C, 2-^{3}H]hexadecanol) have also been used to examine these syntheses that give rise to vinyl ether lipids with cis unsaturation.[71]

Recent work by Snyder and colleagues with cell-free systems isolated from starfish[72] and mouse tumour tissue[73,74] established that microsomes converted dihydroxyacetone phosphate (DHAP) and long-chain fatty alcohols into products containing O-alkyl bonds under suitable incubation conditions. Inhibition studies[74] demonstrated that DHAP was the true triose phosphate precursor. Addition of NADPH was subsequently required for the formation of diacylglyceryl ethers. O-Alkyl-DHAP and a product with the properties of 1-O-alkyl-2-acyl-sn-glycerol 3-phosphate (an analogue of phosphatidate) were identified as intermediates in the synthesis of these lipids.[75] [^{18}O]Hexadecanol was incorporated into the ether oxygen in O-alkyl lipids and the plasmalogen derivatives.[76] On the basis of these results,

Snyder proposed a sequence of reactions for the synthesis of glyceryl ether lipids; these are given in Scheme 5.3. The 1-O-alkyl-2-acylglycerol thus formed may then be converted into alkyl-diacylglycerol or the related phospholipid by reaction with an acyl-CoA or CDP-base respectively,[77] in a process similar to the final stage in the production of triglyceride or phosphatide.

Synthesis of acyl-DHAP was also effected by reaction of DHAP directly with palmitoyl-CoA by guinea-pig mitochondria from various tissues.[78] This product was further converted into the ether, 1-O-alkyl-DHAP, by mitochondrial and microsomal fractions after transfer from a long-chain alcohol.[79] It was subsequently reduced at the C-2 carbonyl group and acylated to form glyceryl ether lipids. Prior incubation of palmitoyl-DHAP and liver mitochondria with NADPH inhibited the formation of alkyl ether and resulted in the reduction of the substrate to 1-acylglycerol 3-phosphate (lysophosphatidate).[80] The precursor-product relationship between acyl and alkyl groups has not really been clarified but a recent report does confirm that acyl-CoA esters act as precursors of the long-chain alkyl groups that are found in the neutral lipid and phosphatide analogues [(IX) and (X)] in neoplasms (Ehrlich ascites cells) in mice.[81]

Biosynthetic Pathways Involving CDP-Diglyceride (CDP-Diacylglycerol)

Kennedy and coworkers have also established the mode of formation of another three classes of lipid, namely phosphoinositides, phosphatidyl-glycerol (and the related cardiolipin) and phosphatidylserine (in bacteria). The mechanism responsible for these syntheses is quite distinct from that determined for the choline and ethanolamine derivatives in that phosphatidate does not undergo dephosphorylation prior to incorporation. CDP-diglyceride plays an essential role and is formed by reaction of phosphatidate with CTP by means of a microsomal cytidylyl transferase, with the release of a pyrophosphate group.[82,83] The concentration of this metabolite in tissues from animal sources is low indicating a high turnover rate. Phosphatidate may also be formed in brain microsomes by means of the direct phosphorylation of a diglyceride with ATP in a reaction catalysed by diglyceride kinase.[83a] This enzyme therefore contributes towards the turnover of the phosphate group after hydrolysis of phosphatidate and could possibly act in providing an additional supply of CDP-diglyceride from an original source other than glycerol phosphate.

With regard to phosphoinositide biosynthesis, the nucleotide substrate acts as acceptor by reacting directly with myo-inositol (under the influence of the microsomal CDP-diglyceride:inositol phosphatidate transferase) with typical cleavage of the pyrophosphate bond[82,84,85] to form 1-phosphatidyl-L-myo-inositol (XI) and CMP (equation (8)):

$$\text{CDP-diglyceride} + \text{inositol} \longrightarrow \text{phosphatidylinositol} + \text{CMP} \qquad (8)$$

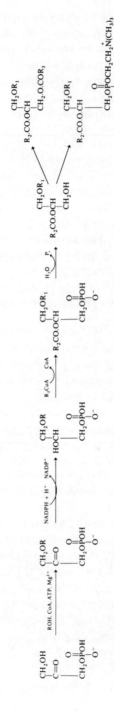

Scheme 5.3   Formation of 1-*O*-alkyl-2-acylglycerol from DHAP and 1-*O*-alkylDHAP

Thus the phosphate group in the product is derived from phosphatidate and hence originally from glycerol phosphate (Scheme 5.4). CDP-diglyceride may also be synthesized in yeasts.[86] However, phosphatidylinositol may apparently be formed from [^{14}C]inositol by particulate preparations in the absence of this liponucleotide.[87] Although synthesis was stimulated to some extent by phosphatidylethanolamine and phosphatidate, they were not themselves incorporated into phosphatidylinositol.

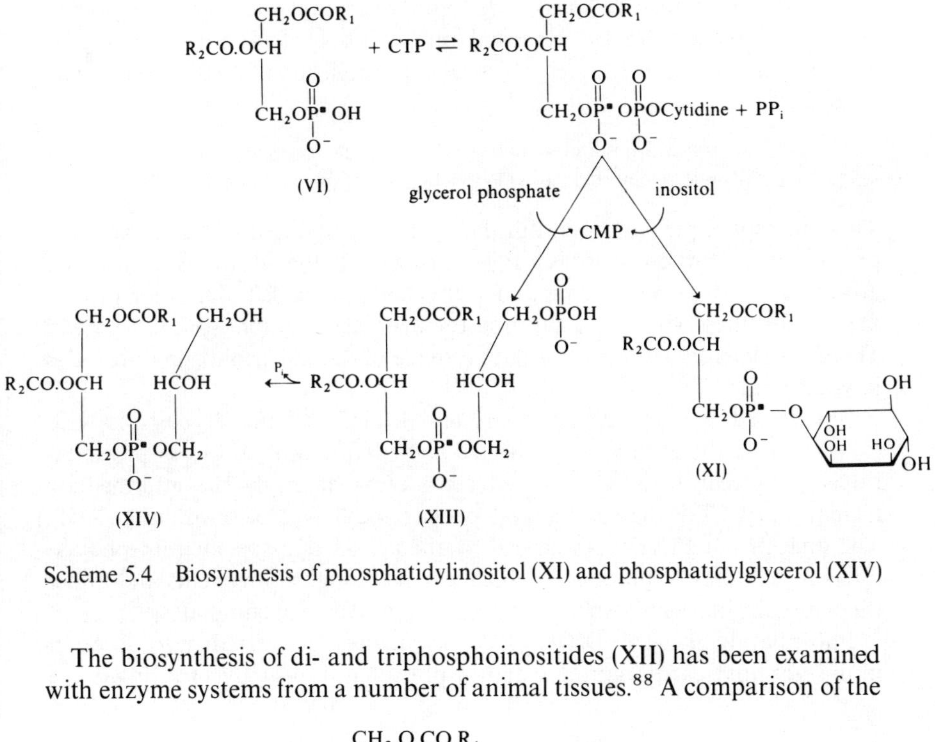

Scheme 5.4 Biosynthesis of phosphatidylinositol (XI) and phosphatidylglycerol (XIV)

The biosynthesis of di- and triphosphoinositides (XII) has been examined with enzyme systems from a number of animal tissues.[88] A comparison of the

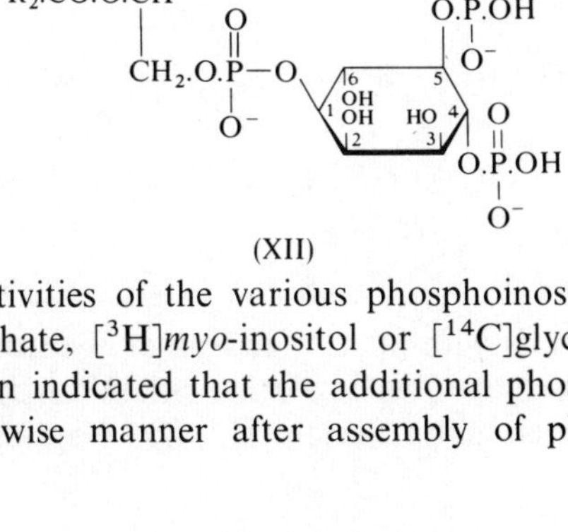

(XII)

relative specific activities of the various phosphoinositides derived from ^{32}P-labelled phosphate, [^{3}H]myo-inositol or [^{14}C]glycerol in subcellular fractions from brain indicated that the additional phosphate groups were inserted in a stepwise manner after assembly of phosphatidylinositol.

Initially, the appropriate enzyme (ATP:phosphatidylinositol 4-phospho-transferase) gives rise to the diphosphoinositide product, 1-phosphatidyl-inositol 4-phosphate. This enzyme is localized in the plasma membrane in brain[89] and liver[90] (as demonstrated by assay of marker enzymes in the various subcellular fractions) and its function may be related to transport of cations, since diphosphoinositide readily complexes with divalent cations. Subsequently the second phosphate group is inserted at the 5-hydroxyl position by means of ATP:diphosphoinositide 5-phosphotransferase to yield triphosphoinositide (XII).[91] In both cases, the additional phosphate group is derived from the terminal phosphate of ATP as evidenced by incorporation of label from $[\gamma\text{-}^{32}P]$ATP and formation of ADP[92] (equations (9) and (10)):

$$\text{Phosphatidylinositol} + \text{ATP} \longrightarrow \text{1,4-diphosphoinositide} + \text{ADP} \qquad (9)$$
$$\text{1,4-Diphosphoinositide} + \text{ATP} \longrightarrow \text{1,4,5-triphosphoinositide} + \text{ADP} \qquad (10)$$

These phospholipids are found in the myelin component of brain and peripheral nerves[93] where there is a rapid turnover of the additional phosphate groups. They also occur in liver and kidney mitochondria[94] and preparations made from these structures catalyse the synthesis of diphosphoinositide.[95] There is evidence in support of the presence of di- and triphosphoinositides in yeast lipids.[96]

Mitochondria from liver,[97] heart and brain,[98,99] and *E. coli* preparations[100] have the ability to synthesize phosphatidylglycerol (3-*sn*-phospha-tidyl-1'-*sn*-glycerol, (XIV)) from glycerol phosphate, *via* the intermediate formation of CDP-diglyceride and phosphatidylglycerol phosphate (XIII). The final product (XIV) is released by the action of a specific phosphatase (Scheme 5.4) that is inhibited by addition of thiol group reagents and under these circumstances phosphatidylglycerol phosphate accumulates.[101]

Phosphatidylglycerol (XIV) may be further converted into another important lipid, cardiolipin [1,3-di(3-*sn*-phosphatidyl)-*sn*-glycerol, (XV)], in a

$$
\begin{array}{l}
\overset{1}{C}H_2.O.CO.R_1 \quad \overset{3'}{C}H_2.O.\overset{O}{\overset{\|}{P}}.O.\overset{3}{C}H_2 \\
\qquad\qquad\qquad\qquad\qquad O^- \\
R_2.CO.O.\overset{2}{C}H \qquad H.\overset{2'}{C}.OH \qquad H\overset{2}{C}.O.CO.R_3 \\
\qquad\qquad O \\
\overset{3}{C}H_2.O.\overset{\|}{P}.O.\overset{1'}{C}H_2 \qquad\qquad \overset{1}{C}H_2.O.CO.R_4 \\
\qquad\quad O^-
\end{array}
$$

(XV)

reaction which proceeds with the elimination of glycerol[102] (in bacteria) or after condensation with a second molecule of CDP-diglyceride and genera-tion of CMP (in mitochondrial systems).[103,103a] Cardiolipin occurs almost exclusively in mitochondrial membranes within the animal kingdom but until recently information concerning its biosynthesis was obtained from particulate preparations of *E. coli*.[102] Reaction takes place between two molecules of phosphatidylglycerol in the absence of CDP-diglyceride although it may be enhanced by its addition to the system. Mitochondria isolated from pig liver are also capable of forming cardiolipin (*via* the intermediacy of CDP-diglyceride and phosphatidylglycerol), when they are provided with glycerol 3-phosphate and CTP as substrates.[103] These metabolites proved accessible to the active mitochondria. Thus the CDP-diglyceride mechanism appears to be operative in animal tissues.

The formation of phosphatidylserine in bacteria is brought about by condensation of CDP-diglyceride with serine.[100] However, this phospholipid is only present in trace amounts in *E. coli* and other bacteria since it is rapidly converted into phosphatidylethanolamine by decarboxylation. Membrane fractions from ghosts of a *Bacillus* sp. (cells with the protoplasm removed, leaving the cell walls and membranes) contain all the enzymes responsible for its biosynthesis from phosphatidate.[104]

BIOSYNTHESIS OF SPHINGOLIPIDS

The detailed mechanisms by which sphingolipids are synthesized has only become clarified in recent years. In the course of the following discussion some mention will be made of an interesting series of inborn errors of sphingolipid metabolism that cause particular lipids to accumulate in certain tissues or cells. Considerable progress has been made in this field and they are now known to arise as a consequence of enzyme deletions in catabolic pathways (with resultant inability to degrade the lipids concerned) rather than biosynthetic pathways. Nevertheless, it is felt that a brief account of these conditions may be justified to gain an overall impression of the importance of the intermediates involved. These are normally present in small amounts and disturbances in their relative concentrations generate far-reaching consequences.

Formation of Sphing-4-enine (Sphingosine)

The early *in vivo* studies in this direction involving labelled substrates originally showed that C-1, C-2 and the amino group of sphing-4-enine (III) were derived from serine[105] while the remaining carbon atoms arose from acetate probably *via* a C_{16} intermediate.[106] The pathway leading to its synthesis was later examined in greater depth by Brady[107,108] and Snell[109] with their colleagues using enzyme preparations from rat brain or yeast microsomes that utilized L-serine, palmitoyl-CoA and NADPH as substrates. Conversion of palmitoyl-CoA into the fully saturated C_{18} product sphinganine (dihydrosphingosine) required Mn^{2+} and involved pyridoxal phosphate-bound serine, suggesting that the reactive form of the acyl sub-

strate might exist as a Schiff base-Mn^{2+}-chelate complex.[110] In the absence
of NADPH, the yeast enzyme did not reduce the carbonyl function and
accordingly the corresponding ketone intermediates, 3-oxo sphinganine
and 3-oxo sphing-4-enine accumulated.[110,111] The presence of these metabolites indicated that a pathway, in which palmitoyl-CoA condensed directly
with pyridoxal phosphate-bound serine, might be operative to give sphinganine as the primary product (equations (11) and (12)):

$$CH_3.[CH_2]_{14}.CO.S.CoA + HO_2C.CH(NH_2).CH_2OH \rightarrow$$

$$CH_3.[CH_2]_{14}.CO.CH(NH_2).CH_2OH + CoA.SH + CO_2 \quad (11)$$

$$CH_3.[CH_2]_{14}.CO.CH(NH_2).CH_2OH + NADPH + H^+ \rightarrow$$

$$CH_3.[CH_2]_{14}.CH(OH).CH(NH_2).CH_2OH + NADP^+ \quad (12)$$

Enzymes catalysing the initial condensation and NADPH-dependent
reduction reactions were isolated from microsomal preparations of yeast
and liver.[109] Palmitaldehyde, a possible precursor of the palmitoyl residue
utilized for sphingolipid formation, was only incorporated into sphinganine
under conditions that permitted conversion into palmitoyl-CoA.[112] It has
not yet been critically established at which stage reduction of the 3-oxo
group or desaturation of the acyl residue takes place, but recent in vivo
studies in which doubly-labelled $[3-^{14}C,3-^3H]$sphinganine was fed to rats
indicated that the isotope ratio remained unchanged in sphing-4-enine.[113]
This result demonstrated that the secondary alcohol group at C-3 was not
affected during this desaturation and therefore favoured initial reduction of
the oxo group. Reaction of sphinganine with a flavoprotein dehydrogenase
may therefore occur as the final step (equation (13)):

$$CH_3[CH_2]_{12}CH_2CH_2CH(OH).CH(NH_2).CH_2OH + flavin \rightarrow$$

$$CH_3[CH_2]_{12}CH:CHCH(OH).CH(NH_2).CH_2OH + flavin.H_2 \quad (13)$$

However, a particulate fraction from the yeast Hansenula ciferri that had
condensing enzyme and reductase activity catalysed the formation of
trans-2-hexadecenoyl-CoA.[114] This product and the saturated palmitoyl-
CoA (but not α-hydroxypalmitoyl-CoA) were good substrates for the condensing enzyme. The α-hydroxy acyl groups are common constituents of the
cerebroside lipids, where they are attached by an amide linkage. The position
and configuration of the double bond in trans-2-hexadecenoyl-CoA is
identical with that ultimately found in sphing-4-enine. Thus there is evidence
suggesting that synthesis of saturated and unsaturated sphingolipids diverges
at the level of acyl-CoA (without the intervention of an additional flavoprotein dehydrogenase) although it is possible, as the authors of this work
point out,[114] that these results may have been complicated by the detergent

properties of the substrates. This view must always be considered when dealing with metabolites of this nature.

Sphingomyelin

The conversion of sphingosine into sphingomyelin involves the acylation of the amino group with a long-chain acyl residue (principally C_{18}, C_{24} and $C_{24:1}$) to give the ceramide product and its subsequent reaction with CDP-choline at the primary alcohol group.[115] This latter reaction is analogous to that which occurs between CDP-choline and the free hydroxyl group in 1,2-diglycerides during the synthesis of lecithin. Enzymes from liver and brain have been described that convert sphingosine plus acyl-CoA[116] or free acid[117] into the amide, N-acylsphingosine (ceramide). The latter enzyme apparently has no requirement for coenzyme A or ATP but functionally it may act catabolically as a hydrolase. Ceramide finally reacts with CDP-choline and the enzyme CDP-choline:ceramide cholinephosphotransferase (EC 2.7.8.3) to yield sphingomyelin (IV),[118] with the release of CMP (equation (14)):

$$CH_3[CH_2]_{12}CH:CHCH(OH)CH(NH)CH_2OH + \text{CDP-choline} \rightleftharpoons$$
$$\overset{|}{\underset{RCO}{}}$$

$$CH_3[CH_2]_{12}CH:CHCH(OH)CH(NH)CH_2O\overset{\overset{O}{\|}}{P}OCH_2CH_2N^+(CH_3)_3 + CMP \quad (14)$$
$$\underset{RCO}{|} \qquad \underset{O^-}{|}$$

$$(IV)$$

An enzyme from chicken liver has been isolated that is active for both the erythro- (natural substrate) and threo-ceramides.[118] It is probable, however, that an additional pathway exists in brain that proceeds *via* reaction of sphingosylphosphorylcholine with a long-chain acyl-CoA and does not utilize a ceramide acceptor as intermediate.[119]

Sphingomyelin accumulates in several tissues of patients suffering with Niemann-Pick disease.[120] This condition is caused by the reduced activity or absence of a sphingomyelin hydrolytic enzyme that cleaves off phosphoryl-choline as catalysed by the choline phosphohydrolase, EC 3.1.4.3) with the formation of ceramide. Some properties of this enzyme isolated from rat tissues have been described which indicate that it is inactive with lecithin and phosphatidylethanolamine as substrate.[121]

Cerebroside

Sphingosine also acts as precursor for other classes of nervous system lipids that lack a phosphodiester group. This is replaced in cerebrosides by a galactose or glucose residue that is present in β-glycosidic linkage. Galactocerebroside is the major lipid component of myelin but glucocere-broside tends to occur in tissues other than brain. The principal steps in its biosynthesis have been described by Brady.[122] Sphingosine reacts with UDP-galactose as donor of the sugar moiety[123] to give psychosine (XVI) and UDP, under the influence of the microsomal enzyme galactosyl-sphingosine transferase. The lipid thus formed undergoes acylation with a long-chain acyl-CoA (mainly C_{24} and $C_{24:1}$[122] and their α-hydroxy-derivatives)[124] to give the cerebroside (V) (equations (15) and (16)):

$$CH_3[CH_2]_{12}CH:CHCH(OH)CH(NH_2)CH_2OH + \text{UDP-galactose} \rightarrow$$

$$-CH(NH_2)CH_2O-\text{galactose} + UDP \quad (15)$$
$$(XVI)$$

$$(V)$$

It was also reported that cerebroside might be synthesized in chicken[125] and mice[126] brain from ceramide and UDP-glucose, that is, in a similar manner to the formation of sphingomyelin. More recently, this pathway has been confirmed with homogenates from late-embryonic chicken brain that utilize UDP-galactose and ceramide (with 2-hydroxystearate as acyl constituent).[127] A second reaction involving ceramide–glucose and UDP-galactose was catalysed by β-galactosyltransferase (isolated from rat spleen) and resulted in ceramide-lactose,[128] a precursor of gangliosides (see below).

In Gaucher's disease glucocerebroside accumulates in the reticulo-endothelial cells of the spleen and also in liver and bone-marrow.[129] The brain[130] and spleen[131] of normal subjects possess a catabolic enzyme (β-D-glucoside glucohydrolase, EC 3.2.1.21) that specifically hydrolyses the glycosidic bond in glucocerebrosides forming glucose and ceramide, but is inactive with galactocerebroside. The greatly decreased activity of this enzyme in the spleen of patients with this condition is responsible for the metabolic lesion that ensues.[132] The glucocerebroside itself is derived by degradation of ceramide–tetrahexoside (globoside), an erythrocyte stromal sphingolipid in which the ceramide unit remains attached to a glucose residue.[133] In the infantile form of this condition, however, the level of glucocerebroside in the brain is also high. Its fatty acid and sphingosine composition reflects that in the ceramide moiety of brain gangliosides and is presumably formed by degradation of these metabolites.

A further modification to the structure of cerebroside is found in the sulphatides[134] in which a sulphate group is esterified at the C-3 position of the galactose residue (XVII). 'Active' sulphate (3'-phosphoadenosine-5'-phosphosulphate, PAPS) is concerned in this sulphotransferase reaction with galactose-containing sphingolipids as substrate.[135] These lipids accumulate in the brain and nervous system, at the expense of unesterified

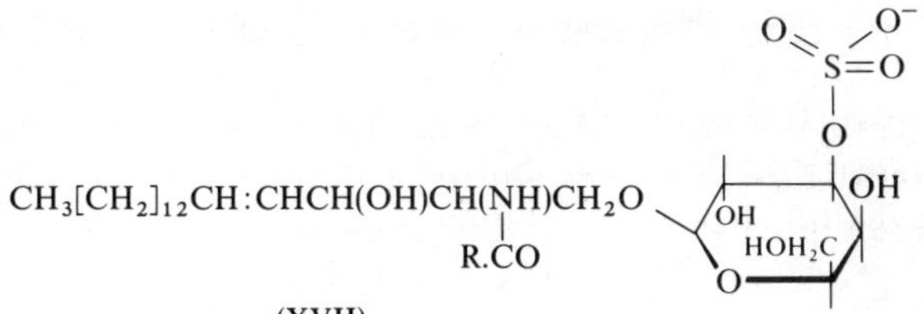

$$CH_3[CH_2]_{12}CH:CHCH(OH)CH(NH)CH_2O$$
$$R.\overset{|}{C}O$$

(XVII)

cerebroside, of individuals with metachromatic leucodystrophy.[136] This condition arises as a consequence of reduced sulphatase activity.

Gangliosides

These metabolites form a rather complex class of sphingolipid (glycolipid) that is characterized by its high carbohydrate content. They are primarily confined to nervous tissue but their synthesis proceeds in many cells concurrently with that of glycoproteins; the carbohydrate portions of both groups of substances are located on cell surfaces. Gangliosides are related structurally and biosynthetically to the cerebrosides and contain an acyl sphingosine (ceramide) group that is attached by its primary hydroxyl group to oligosaccharides of varying chain-length but consisting of glucose, galactose, N-acetylgalactosamine and N-acetylneuraminic acid (NANA,

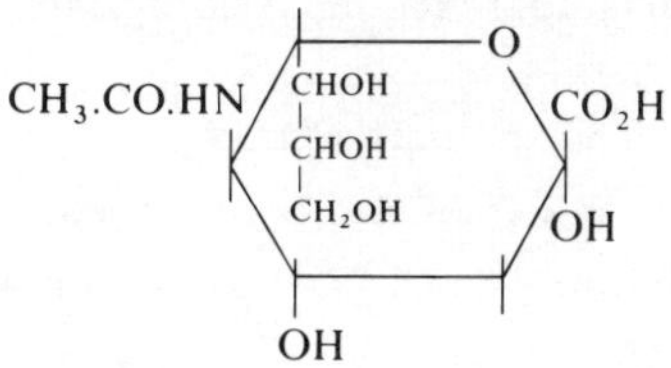

$$CH_3.CO.HN$$

(XVIII)

(XVIII)) residues. The major ganglioside isolated from brain has the structure (XIX) (monosialosyl-N-tetraglycosylceramide, G_{M1}) and is derived

Ceramide ← Glu ← Gal ← N-acGal ← Gal
↑
NANA

(XIX)

from ceramide by glycosylation reactions involving serial transfer of nucleotide-linked sugars using membrane-bound enzymes of high specificity.[137] Similar gangliosides exist in which one or more additional NANA residues (derived from cytidinemonophospho-NANA) are attached to the parent compound.[138] The major fatty acid component is stearic acid (approximately 80–90 per cent)[139] in contrast to the C_{24} acids that are prevalent in cerebrosides.

An accumulation of gangliosides, in particular the monosialoganglioside G_{M2} (XX) gives rise to Tay-Sachs disease. A deficiency of the β-N-acetylgalactosaminidase during the course of degradation of G_{M1} (XIX) which acts

$$\text{Ceramide} \leftarrow \text{Glu} \leftarrow \text{Gal} \leftarrow \textit{N}\text{-acGal}$$
$$\uparrow$$
$$\text{NANA}$$

(XX)

on the Gal $\leftarrow$ N-acGal sequence may be responsible for the build-up of G_{M2} gangliosides.[140] A possible reason for this condition related to lack of the UDP-galactose:glycolipid galactosyltransferase enzyme seems unlikely since incorporation of UDP-[^{14}C]galactose is not impaired in brain tissue from patients suffering from this disease.[140]

GALACTOLIPIDS

Finally, a brief mention should be made of the group of galactolipids that is very abundant in plant tissues. Monogalactosyl diglyceride (XXI) and the digalactosyl derivative are the major types with α-linolenic acid ($C_{18:3,\Delta^{9,12,15}}$) as the principal fatty acid component. They act as structural

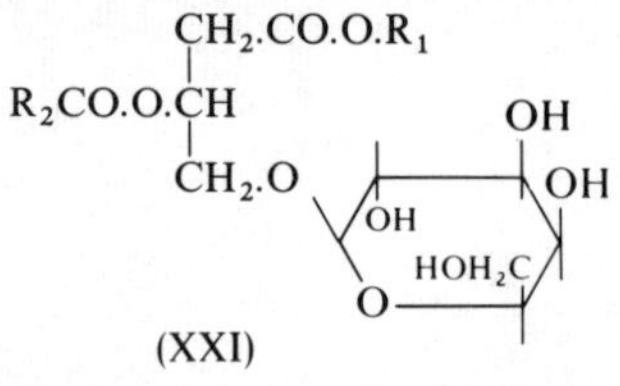

$$CH_2.CO.O.R_1$$
$$R_2CO.O.CH$$
$$CH_2.O$$

(XXI)

components in the lamellar membrane within the chloroplast and are formed by reaction of suitable diglyceride acceptors with UDP-galactose. Different substrates and enzymes appear to be required for the synthesis of the mono- and disaccharide derivatives.[141] A soluble transferase enzyme has recently been obtained from sub-chloroplast fractions of spinach leaves.[142]

INTERRELATIONSHIPS BETWEEN THE MAJOR CLASSES OF LIPIDS

To bring this chapter to a close, the biosynthetic interrelationships between the different routes leading to the various lipid products are summarized in Scheme 5.5. This presentation attempts to show the participation of common intermediates.

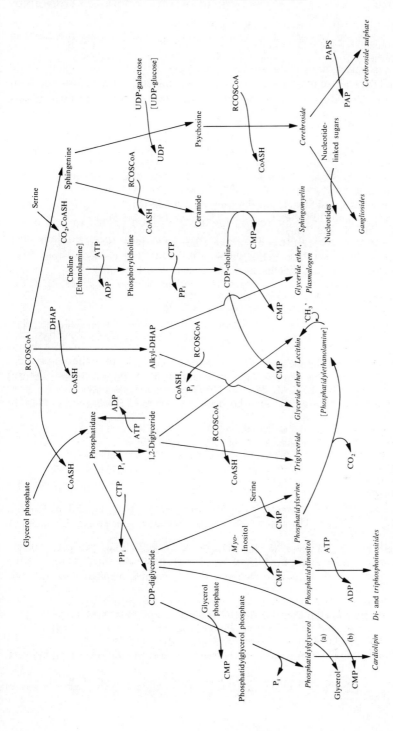

Scheme 5.5   Summary of reaction sequences involved in the biosynthesis of lipids

REFERENCES

1. Danielli, J. F., and Davson, H., *J. Cell. Comp. Physiol.*, **5**, 495 (1935)
2. Finean, J. B., and Robertson, J. D., *Brit. Med. Bull.*, **14**, 267 (1958)
3. Vandenheuval, F. A., *Ann. N.Y. Acad. Sci.*, **122**, 57 (1965)
4. Rothfield, L., and Finkelstein, A., *Annu. Rev. Biochem.*, **37**, 463 (1968)
5. In *IUPAC-IUB Commission on Biochemical Nomenclature: The Nomenclature of Lipids, Biochem. J.*, **105**, 897 (1967)
6. Kennedy, E. P., *Fed. Proc.*, **20**, 934 (1961)
7. Stein, Y., Tietz, A., and Shapiro, B., *Biochim. Biophys. Acta*, **26**, 286 (1957)
8. Weiss, S. B., Kennedy, E. P., and Kiyasu, J. Y., *J. Biol. Chem.*, **235**, 40 (1960)
9. Stein, Y., and Shapiro, B., *Biochim. Biophys. Acta*, **30**, 271 (1958)
10. Wilgram, G. F., and Kennedy, E. P., *J. Biol. Chem.*, **238**, 2615 (1963)
11. Bublitz, C., and Kennedy, E. P., *J.Biol. Chem.*, **211**, 951 (1954)
12. Wieland, O., and Suyter, M., *Biochem. Z.*, **329**, 320 (1957)
13. Clark, B., and Hübscher, G., *Nature (London)*, **195**, 599 (1962)
14. Margolis, S., and Vaughan, M., *J. Biol. Chem.*, **237**, 44 (1962)
15. Lands, W. E. M., and Hart, P., *J. Biol. Chem.*, **240**, 1905 (1965)
16. Lamb, R. G., and Fallon, H. J., *J. Biol. Chem.*, **245**, 3075 (1970)
17. Mårtensson, E., and Kanfer, J., *J. Biol. Chem.*, **243**, 497 (1968)
18. Van den Bosch, H., and Vagelos, P. R., *Biochim. Biophys. Acta*, **218**, 233 (1970)
19. Goldfine, H., *J. Biol. Chem.*, **241**, 3864 (1966)
20. Kuhn, N. J., and Lynen, F., *Biochem. J.*, **94**, 240 (1965)
21. Smith, S. W., Weiss, W. B., and Kennedy, E. P., *J. Biol. Chem.*, **228**, 915 (1957)
22. Hanahan, D. J., Brockerhoff, H., and Barron, E. J., *J. Biol. Chem.*, **235**, 1917 (1960)
23. Savary, P., and Desnuelle, P., *Biochim. Biophys. Acta*, **50**, 319 (1961)
24. Possmayer, F., Scherphof, G. L., Dubbelman, T. M. A. R., van Golde, L. M. G., and van Deenen, L. L. M., *Biochim. Biophys. Acta* **176**, 95 (1969)
25. Elovson, J., Åkesson, B., and Arvidson, G., *Biochim. Biophys. Acta*, **176**, 214 (1969)
26. Brockerhoff, H., *J. Lipid Res.*, **6**, 10 (1965)
27. Slakey, P. M., and Lands, W. E. M., *Lipids*, **3**, 30 (1968)
28. Lands, W. E. M., and Merkl, I., *J. Biol. Chem.*, **238**, 898, 905 (1963)
29. Hill, E. E., and Lands, W. E. M., *Biochim. Biophys. Acta*, **152**, 645 (1968)
30. Erbland, J. F., Brossard, M., and Marinetti, G. V., *Biochim. Biophys. Acta*, **137**, 23 (1967)
31. Marinetti, G. V. In *Comprehensive Biochemistry*, Vol. 18, p. 117. Ed. by Florkin, M., and Stotz, E. H., Elsevier Publishing Company, Amsterdam, London and New York, 1970
32. Steinberg, D., Vaughan, M., and Margolis, S., *J. Biol. Chem.*, **236**, 1631 (1961)
33. Tzur, R., and Shapiro, B., *J. Lipid Res.*, **5**, 542 (1964)
33a. Roncari, D. A. K., and Hollenberg, C. H., *Biochim. Biophys. Acta*, **137**, 446 (1967)
34. Goldman, P., and Vagelos, P. R., *J. Biol. Chem.*, **236**, 2620 (1961)
35. Clark, B., and Hübscher, G., *Biochim. Biophys. Acta*, **46**, 479 (1961)
36. Brown, J. L., and Johnston, J. M., *Biochim. Biophys. Acta*, **84**, 448 (1964)
37. Paris, R., and Clement, G., *Biochim. Biophys. Acta*, **152**, 63 (1968)
38. Johnston, J. M., Rao, G. A., and Lowe, P. A., *Biochim. Biophys. Acta*, **137**, 578 (1967)
39. Rao, G. A., and Johnston, J. M., *Biochim. Biophys. Acta*, **144**, 25 (1967)
40. Ailhaud, G. P., and Vagelos, P. R., *J. Biol. Chem.*, **241**, 3866 (1966)

41. Kuhn, N. J., *Biochem. J.*, **105**, 225 (1967)
42. Flatt, J. P., *J. Lipid Res.*, **11**, 131 (1970)
43. Tepperman, J., and Tepperman, H. M., *Ann. N.Y. Acad. Sci.*, **131**, 404 (1965)
44. Robinson, D. S. In *Comprehensive Biochemistry*, Vol. 18, p. 51. Ed. by Florkin, M., and Stotz, E. H. Elsevier Publishing Company, Amsterdam, London and New York, 1970
45. Del Boca, J., and Flatt, J. P., *Eur. J. Biochem.*, **11**, 127 (1969)
46. Kornberg, A., and Pricer, W. E., *Fed. Proc.*, **11**, 242 (1952)
47. Kennedy, E. P., and Weiss, S. B., *J. Biol. Chem.*, **222**, 193 (1956)
48. Strickland, K. P., Subrahmanyam, D., Pritchard, E. T., Thompson, W., and Rossiter, R. J. *Biochem. J.*, **87**, 128 (1963)
49. Borkenhagen, L. F., and Kennedy, E. P., *J. Biol. Chem.*, **227**, 951 (1957)
50. Weiss, S. B., Smith, S. W., and Kennedy, E. P., *J. Biol. Chem.*, **231**, 53 (1958)
51. Bjørnstad, P., and Bremer, J., *J. Lipid Res.*, **7**, 38 (1966)
52. Dawson, R. M. C. In *Essays in Biochemistry*, Vol. 2, p. 69. Ed. by Campbell, P. N., and Greville, G. D., Academic Press Inc., London and New York, 1966
53. Hübscher, G. *Biochim. Biophys. Acta*, **57**, 555 (1962)
54. Rehbinder, D., and Greenberg, D. M., *Arch. Biochem. Biophys.*, **109**, 110 (1965)
55. Wilgram, G. F., and Kennedy, E. P., *J. Biol. Chem.*, **238**, 2615 (1963)
56. McMurray, W. C., and Dawson, R. M. C., *Biochem. J.*, **112**, 91 (1967)
57. Jungalwala, F. B., and Dawson, R. M. C., *Eur. J. Biochem.*, **12**, 399 (1970)
58. Jungalwala, F. B., and Dawson, R. M. C., *Biochem. J.*, **117**, 481 (1970)
59. Wirtz, K. W. A., and Zilversmit, D. B., *Biochim. Biophys. Acta*, **193**, 105 (1969)
60. Lands, W. E. M., Blank, M. L., Nutter, L. J., and Privett, O. S., *Lipids*, **1**, 224 (1966)
61. Shephard, E. H., and Hübscher, G., *Biochem. J.*, **113**, 429 (1969)
62. Galliard, T., Michell, R. H., and Hawthorne, J. N., *Biochim. Biophys. Acta*, **106**, 551 (1965)
63. Hajra, A. K., Seguin, E. B., and Agranoff, B. W., *J. Biol. Chem.*, **243**, 1609 (1968)
64. Sribney, M., *Arch. Biochem. Biophys,*. **126**, 954 (1968)
65. Thompson, G. A. In *Comprehensive Biochemistry*, Vol. 18, p. 157. Ed. by Florkin, M., and Stotz, E. H. Elsevier Publishing Company, Amsterdam, London and New York, 1970
66. Kiyasu, J. Y., and Kennedy, E. P., *J. Biol. Chem.*, **235**, 2590 (1960)
67. Thompson, G. A., *J. Biol. Chem.*, **240**, 1912 (1965)
68. Ellingboe, J., and Karnovsky, M. L., *J. Biol. Chem.*, **242**, 5693 (1967)
69. Thompson, G. A. *Biochim. Biophys. Acta*, **152**, 409 (1968)
70. Wood, R., and Healy, K., *Biochem. Biophys. Res. Commun.*, **38**, 205 (1970)
71. Stoffel, W., and Lekim, D., *Hoppe-Seyler's Z. Physiol. Chem.*, **352**, 501 (1971)
72. Snyder, F., Malone, B., and Blank, M. L., *Biochim. Biophys. Acta*, **187**, 302 (1969)
73. Anderson, R. E., Cumming, R. B., Walton, M., and Snyder, F., *Biochim. Biophys. Acta*, **176**, 491 (1969)
74. Snyder, F., Malone, B., and Blank, M. L., *J. Biol. Chem.*, **245**, 1790 (1970)
75. Snyder, F., Blank, M. L., Malone, B., and Wykle, R. L. *J. Biol. Chem.*, **245**, 1800 (1970)
76. Snyder, F., Rainey, W. T., Blank, M. L., and Christie, W. H., *J. Biol. Chem.*, **245**, 5853 (1970)
77. Snyder, F., Blank, M. L., and Malone, B., *J. Biol. Chem.*, **245**, 4016 (1970)
78. Hajra, A. K., *J. Biol. Chem.*, **243**, 3458 (1968)
79. Hajra, A. K., *Biochem. Biophys. Res. Commun.*, **39**, 1037 (1970)

80. Hajra, A. K., and Agranoff, B. W., *J. Biol. Chem.* **243,** 3542 (1968)
81. Wood, R., and Healy, K., *J. Biol. Chem.*, **245,** 2640 (1970)
82. Paulus, H., and Kennedy, E. P., *J. Biol. Chem.*, **235,** 1303 (1960)
83. Carter, J. R., and Kennedy, E. P., *J. Lipid Res.*, **7,** 678 (1966)
83a. Hokin, M. R., and Hokin, L. E., *J. Biol. Chem.*, **234,** 1381 (1959)
84. Thompson, W., Strickland, K. P., and Rossiter, R. J., *Biochem. J.*, **87,** 136 (1963)
85. Benjamins, J. A., and Agranoff, B. W., *J. Neurochem.*, **16,** 513 (1967)
86. Hutchison, H. T., and Cronan, J. E., *Biochim. Biophys. Acta*, **164,** 606 (1968)
87. White, G. L., and Hawthorne, J. N., *Biochem. J.*, **117,** 203 (1970)
88. Brockerhoff, H., and Ballou, C. E., *J. Biol. Chem.*, **237,** 49, 1764 (1962)
89. Kai, M., White, G. L., and Hawthorne, J. N., *Biochem. J.*, **101,** 328 (1966)
90. Michell, R. H., Harwood, J. L., Coleman, R., and Hawthorne, J. N., *Biochim. Biophys. Acta*, **144,** 649 (1967)
91. Salway, J. G., Harwood, J. L., Kai, M., White, G. L., and Hawthorne, J. N., *J. Neurochem.*, **15,** 221 (1968)
92. Eichberg, J., and Dawson, R. M. C., *Biochem. J.*, **96,** 644 (1965)
93. Sheltawy, A., and Dawson, R. M. C., *Biochem. J.*, **100,** 12 (1966)
94. Galliard, T., and Hawthorne, J. N., *Biochim. Biophys. Acta*, **70,** 479 (1963)
95. Garbus, J., Deluca, H. F., Loomans, M. E., and Strong, F. M., *J. Biol. Chem.*, **238,** 59 (1963)
96. Lester, R. L., and Steiner, M. R., *J. Biol. Chem.*, **243,** 4889 (1968)
97. Kiyasu, J. Y., Pieringer, R. A., Paulus, H., and Kennedy, E. P., *J. Biol. Chem.*, **238,** 2293 (1963)
98. Stanacev, N. Z., Stuhne-Sekalec, L., Brookes, K. B., and Davidson, J. B., *Biochim. Biophys. Acta*, **176,** 650 (1969)
99. Davidson, J. B., and Stanacev, N. Z., *Can. J. Biochem.*, **48,** 633 (1970)
100. Kanfer, J., and Kennedy, E. P., *J. Biol. Chem.*, **239,** 1720 (1964)
101. Chang, Y., and Kennedy, E. P., *J. Lipid Res.*, **8,** 447, 456 (1967)
102. Hirschberg, C. B., and Kennedy, E. P., *Proc. Nat. Acad. Sci. U.S.*, **69,** 648 (1972)
103. Davidson, J. B., and Stanacev, N. Z., *Biochem. Biophys. Res. Commun.* **42,** 1191 (1971)
103a. Stanacev, N. Z., Davidson, J. B., Stuhne-Sekalec, L., and Domazet, Z., *Biochem. Biophys. Res. Commun.*, **47,** 1972 (1971)
104. Patterson, P. H., and Lennarz, W. J., *J. Biol. Chem.*, **246,** 1062 (1971)
105. Sprinson, D. B., and Coulon, A., *J. Biol. Chem.*, **207,** 585 (1954)
106. Zabin, I., and Mead, J. F., *J. Biol. Chem.*, **205,** 271 (1953)
107. Brady, R. O., and Koval, G. J., *J. Biol. Chem.*, **233,** 26 (1958)
108. Brady, R. O., Formica, J. V., and Koval, G. J., *J. Biol. Chem.*, **233,** 1072 (1958)
109. Braun, P. E., and Snell, E. E., *Proc. Nat. Acad. Sci. U.S.*, **58,** 298 (1967)
110. Braun, P. E., and Snell, E. E., *J. Biol. Chem.*, **243,** 3775 (1968)
111. Brady, R. N., di Mari, S. J., and Snell, E. E., *J. Biol. Chem.*, **244,** 491 (1969)
112. Braun, P. E., Morell, P., and Radin, N. S., *J. Biol. Chem.*, **245,** 335 (1970)
113. Stoffel, W., Assmann, G., and Bister, K., *Hoppe-Seyler's Z. Physiol. Chem.*, **352,** 1531 (1971)
114. di Mari, S. J., Brady, R. N., and Snell, E. E., *Arch. Biochem. Biophys.*, **143,** 553 (1971)
115. Sribney, M., and Kennedy, E. P., *J. Biol. Chem.*, **233,** 1315 (1958)
116. Sribney, M., *Biochim. Biophys. Acta*, **125,** 542 (1966)
117. Gatt, S., *J. Biol. Chem.*, **241,** 3724 (1966)
118. Fujino, Y., Nakano, M., Negishi, T., and Ito, S., *J. Biol. Chem.* **243,** 4650 (1968)
119. Fujino, Y., and Negishi, T., *Biochim. Biophys. Acta*, **152,** 428 (1968)

120. Brady, R. O., Kanfer, J. N., Mock, M. B., and Fredrickson, D. S., *Proc. Nat. Acad. Sci. U.S.*, **55**, 366 (1966)
121. Kanfer, J. N., Young, O. M., Shapiro, D., and Brady, R. O., *J. Biol. Chem.*, **241**, 1081 (1966)
122. Brady, R. O., *J. Biol. Chem.*, **237**, PC2416 (1962)
123. Cleland, W. W., and Kennedy, E. P., *J. Biol. Chem.*, **235**, 45 (1960)
124. Kishimoto, Y., and Radin, N. S., *J. Lipid Res.*, **1**, 72, 79 (1959)
125. Basu, S., Kaufman, B., and Roseman, S., *J. Biol. Chem.*, **243**, 5802 (1968)
126. Morell, P., and Radin, N. S., *Biochemistry*, **8**, 506 (1969)
127. Basu, S., Schultz, A. M., Basu, M., and Roseman, S., *J. Biol. Chem.*, **246**, 4272 (1971)
128. Hauser, G., *Biochem. Biophys. Res. Commun.*, **28**, 502 (1967)
129. Rosenberg, A., and Chargaff, E., *J. Biol. Chem.*, **233**, 1323 (1958)
130. Gatt, S., and Rapport, M. M. *Biochim. Biophys. Acta*, **113**, 567 (1966)
131. Brady, R. O., Kanfer, J. N., and Shapiro, D., *J. Biol. Chem.*, **240**, 39 (1965)
132. Brady, R. O., Kanfer, J. N., and Shapiro, D., *Biochem. Biophys. Res. Commun.*, **18**, 221 (1965)
133. Yamakawa, T., Yokoyama, S., and Handa, N., *J. Biochem. (Tokyo)*, **53**, 28 (1963)
134. Stoffyn, P., and Stoffyn, A., *Biochim. Biophys. Acta*, **70**, 218 (1963)
135. Cumar, F. A., Barra, H. S., Maccioni, H. J., and Caputto, R., *J. Biol. Chem.*, **243**, 3807 (1968)
136. Mehl, E., and Jatzkewitz, H., *Biochem. Biophys. Res. Commun.*, **19**, 407 (1965)
137. Kaufman, B., Basu, S., and Roseman, S., *J. Biol. Chem.*, **243**, 5804 (1968)
138. Svennerholm, L., *J. Lipid Res.*, **5**, 145 (1964)
139. Trams, E. G., Giuffrida, L. E., and Karmen, A., *Nature (London)*, **193**, 680 (1962)
140. Svennerholm, L., *Biochem. J.*, **111**, 6P (1968)
141. Ongun, A., and Mudd, J. B., *J. Biol. Chem.*, **243**, 1558 (1968)
142. Chang, S. B., and Kulkarni, N. D., *Phytochemistry*, **9**, 927 (1970)

CHAPTER 6

Biosynthesis of Phenols and Other Polyketides

Polyketides form a large class of natural product that possess structures of great diversity but are related by their biogenetic origins. Fatty acids, single and multi-ringed phenols, and polyacetylenes are included in this general classification as they are all assembled from C_2 units derived from acetate. Fatty acids occur universally in Nature and their mode of synthesis has been fully discussed in Chapters 2 and 3 of this text. With the exception of the terminal carboxyl group, the original carbonyl functions are all completely reduced. Other classes of polyketides, as their name really implies, usually undergo little or no reduction in the course of their synthesis and retain many or all the carbonyl groups, often as phenolic hydroxyl groups. Another major type of chemically defined polyketide includes the polyacetylenes which are found in plants and fungi, and shows a distinct resemblance to the fatty acids. Indeed there is evidence in certain cases indicating that acetylenic metabolites with a C_{18} chain-length are derived from the related fatty acid.[1]

A number of important features are common to the polyketides other than fatty acids. They have restricted taxonomic distribution and are only formed under certain circumstances; many are water-soluble and are secreted into the culture medium. These compounds are mainly confined to fungi, lichens, actinomycetes and higher plants and have been classified as 'shunt' or 'secondary' metabolites. Primary metabolites, by distinction, are those substances involved in the widely distributed pathways of major importance in all organisms. Many polyketides are frequently formed in high yield usually at a late stage in the growth cycle. They often possess antibiotic activity or other physiological properties that may possibly confer some benefit on the producing organism (although there is little evidence to support this suggestion) but they have certainly aided human society. These properties were sufficient to stimulate interest and the products have presented challenging problems in the elucidation of their structure and pathway of biosynthesis. A model is proposed by the Author at the end of this chapter that attempts to explain the nature and origin of the enzyme complex which is responsible for the synthesis of acetate-derived phenols in fungi.

112

BIOSYNTHESIS OF PHENOLS

Early Developments

One of the two principal routes leading to the biosynthesis of phenols is based on acetyl-CoA and malonyl-CoA condensation reactions (acetate pathway); the other depends on the intermediacy of the non-aromatic metabolites, dehydroquinic, shikimic and chorismic acids (shikimic acid pathway) that derive from phosphoenolpyruvate and erythrose 4-phosphate. With regard to the former route with which this chapter is partly concerned, originally Collie[2] and later Robinson[3] recognized that the synthesis of phenols might involve head to tail linkage of acetate units, whereby the carboxyl group of one molecule reacted with the methyl group of another. However, the first definitive proposals were made by Birch and Donovan[4] whose studies were based on the known importance of acetate as precursor of fatty acids and cholesterol: They noted the correlation between 'marker' oxygen functions in many phenols, for example, acyl phloroglucinol and orcinol structures, and their probable derivation from the carboxyl group of acetate. This supported the hypothesis that these natural products were elaborated by head to tail condensation reactions, followed by intramolecular cyclization *via* C-acylation or aldol condensation. Variations of this type are illustrated in Scheme 6.1. Accordingly a C_8 chain (3,5,7-trioxooctanoic acid) could give rise to structures (a) or (b), where R represents a methyl group, by different cyclization patterns. A single reductive step and elimination of a molecule of water from the hydroxy acyl intermediate at the aliphatic stage

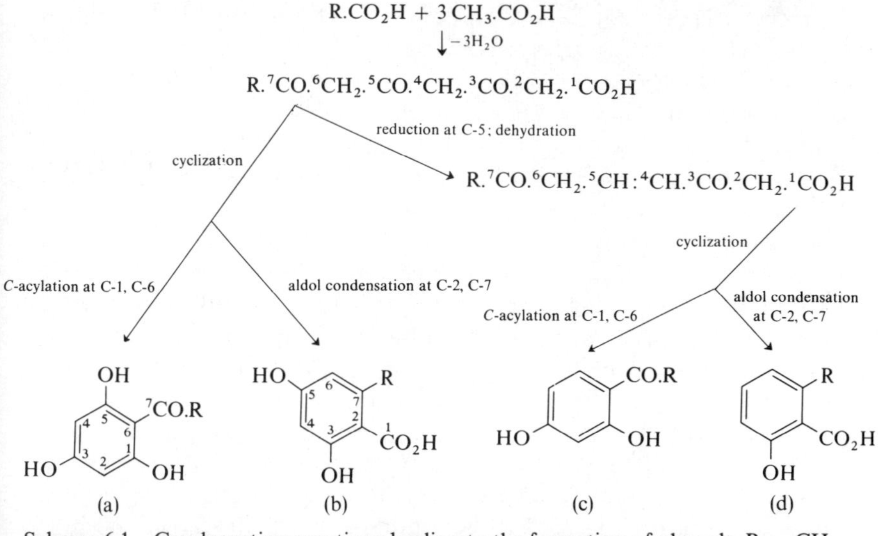

$$R.CO_2H + 3\,CH_3.CO_2H$$
$$\downarrow -3H_2O$$
$$R.^7CO.^6CH_2.^5CO.^4CH_2.^3CO.^2CH_2.^1CO_2H$$

reduction at C-5; dehydration

cyclization

$$R.^7CO.^6CH_2.^5CH:^4CH.^3CO.^2CH_2.^1CO_2H$$

cyclization

C-acylation at C-1, C-6

aldol condensation at C-2, C-7

C-acylation at C-1, C-6

aldol condensation at C-2, C-7

(a)   (b)   (c)   (d)

Scheme 6.1 Condensation reactions leading to the formation of phenols. R = CH_3— or other suitable group, e.g. $C_6H_5.CH:CH$— (after Birch and Donovan, 1953)[4] would give rise to other commonly occurring phenols of the acyl resorcinol (c) or *m*-cresol (d) type.

Degradative studies established the general validity of this theory after [^{14}C]acetate was initially employed as labelled substrate. Many fungal metabolites gave the required distribution pattern and were indeed derived directly from acetate. The examples 6-methylsalicylic acid (I),[5] griseofulvin (II),[6] javanicin,[7] a dihydroxynaphthaquinone derivative (III) and alternariol (IV)[8] have been selected for illustration in Scheme 6.2 and show the probable folding patterns and sites of incorporation of [1-^{14}C]acetate into the expected positions. Metabolites (II), (III) and (IV) all derive (the latter exclusively) from 7 molecules of acetate by different folding arrangements. It is interesting

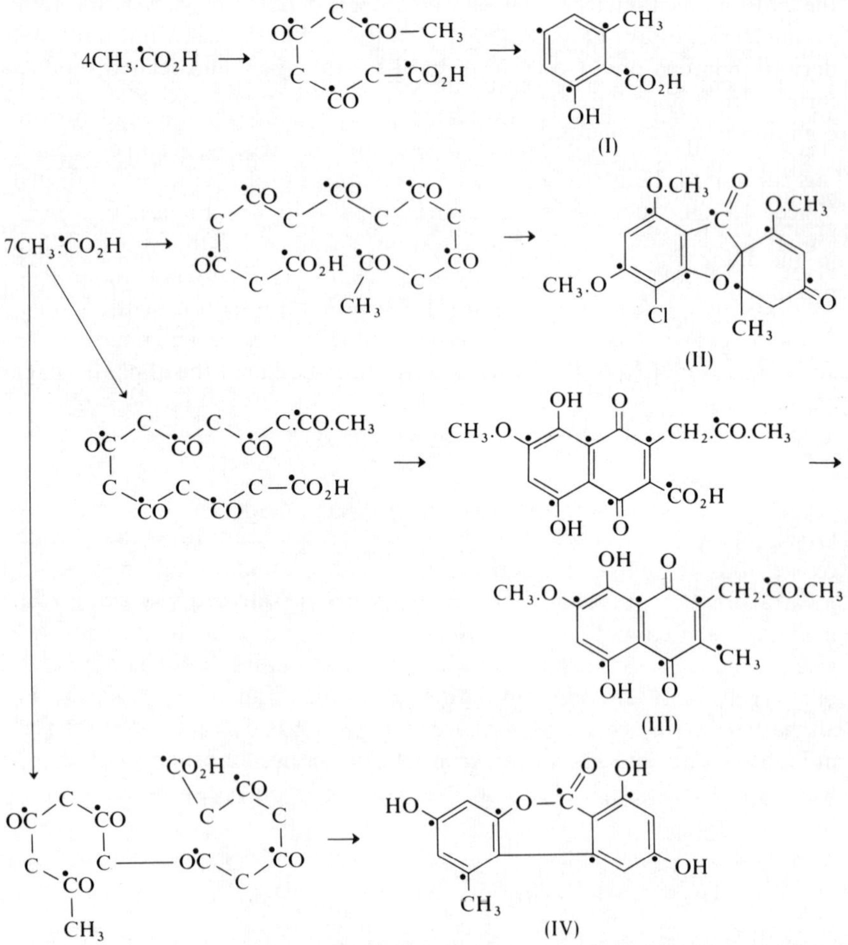

Scheme 6.2   Incorporation of [1-^{14}C]acetate into phenolic metabolites in fungi. The labelled carbon atoms are marked

to note that the ring methyl group of javanicin arises from the carboxyl group of acetate after reduction.[7] [2-^{14}4]Acetate yields similar incorporations into the adjacent carbon atoms but the results are rather more complicated in this instance as the label becomes randomized after passage through the tricarboxylic acid cycle at the level of succinate, with subsequent formation of oxaloacetate, pyruvate and [1-^{14}C]acetyl-CoA. Thus the acetate theory, as reinstated by Birch and his collaborators, and its confirmation began to place the multiplicity of phenolic products belonging to various ring systems into a biogenetic perspective.

Many more complex examples of acetate-derived compounds have since been recorded including those with different initiating groups. Probably the most important from the pharmacological point of view is the tetra-cycline series of antibiotics that may utilize a malonamoyl unit (an amide derived from malonyl-CoA) as starter. Cinnamic acid and related products formed from phenylalanine and tyrosine or their precursors are especially common in higher plants.

Studies on the biosynthesis of orsellinic acid (V) with ^{18}O-labelled acetate confirmed that the ^{18}O was incorporated into the hydroxyl and carboxyl groups but the dilution found for the latter group was twice as great as that in the hydroxyl group.[9] Hydrolysis of the aromatic acyl thioester that presumably exists as intermediate with $H_2{}^{16}O$ would introduce the second oxygen atom from the medium in the formation of the free acid (V), as shown in equation (1):

$$CH_3.\overset{\bullet}{C}\overset{\bullet}{O}OH \rightarrow 4\,CH_3.\overset{\bullet}{C}O.S.CoA \rightarrow CH_3.\overset{\bullet}{C}O.[CH_2.\overset{\bullet}{C}O]_2.CH_2.\overset{\bullet}{C}O.S.X \xrightarrow{H_2O} - - \rightarrow \quad (1)$$

(V)

Orsellinic acid is formed solely from four acetate units with the phenolic oxygen atoms retained; it is distributed widely among fungi. Synthesis of most aromatic compounds, however, involves modifications of the inter-mediate polyketomethylene derivative before and/or after cyclization to give a wide range of metabolites with diverse structure. Reduction (removal of oxygen), introduction of additional oxygen functions, halogenation, alkylation or ring cleavage may occur. Flavonoids characteristically occur in higher plants and are formed from a combination of primary precursors. Ring A of the flavanol, quercitin (VI), is derived from 3 molecules of acetate

(VI)

while ring B, together with the adjoining C_3 unit, is formed from a $C_6.C_3$ intermediate *via* the shikimic acid pathway.[10] The flavonoid pigments are found most frequently in flowers and fruit and usually exist in glycosidic linkage with sugar components.

Involvement of Malonyl-CoA

Following this brief summary of the nature of various polyketides, we now return to a discussion of their mode of biosynthesis. Lynen[11] first drew attention to the possibility, considered on thermodynamic grounds, that malonyl-CoA might be the true condensing agent in the assembly process rather than acetyl-CoA in a manner similar to that engaged in the formation of fatty acids. Decarboxylation of the malonyl residue accompanying condensation would encourage the formation of β-oxoacyl groups. These substances, e.g. 6-methylsalicylic acid (I),[12-14] orsellinic acid (V)[15] and alternariol (IV)[16] are, in fact, synthesized from 1 molecule of acetyl-CoA plus malonyl-CoA. The C-methyl carbon atoms in these and similar products are derived from the acetyl-CoA starter unit which primes the sequence. The tetracyclines are also formed from malonyl groups[17] with malonyl-CoA or malonamoyl-CoA, its amide derivative, as primer. It would seem highly probable that all phenolic compounds formed from acetate (without the intervention of mevalonate) arise in this way. A significant feature of this process is that the growing aliphatic chain undergoes limited or no reduction in contradistinction to the situation recognized for fatty acid synthesis. The presence of the poly-β-ketone structure, however, would confer great reactivity on the intermediates which must, therefore, be stabilized prior to cyclization to permit chain extension. This situation could be achieved after enolization and maintenance of the resultant structure in fixed stereospecific conformation (dependent on the nature of the individual enzyme) by binding of the enolate anions to various amino acid sites, possibly assisted by metal ion chelation.[18] It is also possible that the oxygen functions may be held in similar rigid positions by means of hydrogen bonding; in either case spatial restrictions would be imposed upon the molecule.

Failure to detect free polyketomethylene derivatives in the synthesis of aromatic products supports the concept of an enzyme-bound series of reactions. However, stabilized ring compounds consisting of three acetate residues have been isolated from culture media of *Penicillia*. Triacetic acid lactone (VII),[19] containing an oxygen heterocyclic ring system, is secreted

(VII)

(VIII)

alongside the C_8 product, 6-methylsalicylic acid (I), by *P. patulum* while the
3-methyl derivative (3,6-dimethyl-4-hydroxypyran-2-one, (VIII))[20] is produced by *P. stipitatum* together with the C_9 tropolones (compounds containing a 7-membered ring system). Triacetic acid lactone may also be
formed by fatty acid synthetases from acetyl-CoA plus 2 molecules of
malonyl-CoA when deprived of NADPH (Chapter 4, p. 74), directly confirming their enzymic capability for condensation reactions entailing acetoacetyl and malonyl residues. The lactone is released after nucleophilic attack
by the enolate anion on the thioester carbon atom;[21] this feature is discussed in greater detail later in the chapter (Scheme 6.8). Tetraacetic acid
lactone (IX) has also been found in media from ethionine-inhibited cultures

$$CH_3.CO.CH_2$$
(IX)

of *P. stipitatum* in which synthesis of the tropolones was greatly hindered,
and isomerizes slowly under mild conditions into orsellinic acid (V).[22]

Cell-free Studies on the Biosynthesis of Phenols

Until fairly recently, research into the more biochemical aspects of this
area of metabolism had been neglected. Light and Lynen, however, have
independently applied a more direct approach towards an understanding of
the nature of the enzyme systems involved. Attempts at the preparation of
cell-free extracts from fungi with aromatic synthetic capacity have generally
centred around *P. patulum* (*P. urticae* appears to be the same species), an
organism that secretes 6-methylsalicylic acid and derived substances. The
first such attempt was made over a decade ago by Lynen and Tada[12] who
obtained a soluble extract that was effective in synthesizing this metabolite
from acetyl-CoA and malonyl-CoA in the presence of NADPH. A close
analogy with fatty acid synthesis was therefore apparent which led Lynen[23]
to speculate that a multienzyme complex, to which the intermediates would
be bound by thioester linkage, might be implicated.

Light[14] extended this work and succeeded in gaining more active preparations from frozen mycelia by grinding with sand in buffers of high ionic
strength. 6-Methylsalicylate synthetase activity was associated with the
non-membranous portion of the cells. Purification of the soluble enzyme
was effected by centrifugation but further treatment of the original supernatant resulted in loss of activity. Despite this behaviour he demonstrated
that it possessed a high molecular weight by procedures involving gel filtration and sucrose density centrifugation,[24] supporting its probable identity
as a multienzyme complex.

The effect of addition of inhibitors of protein synthesis to growing cultures of *P. patulum* was also tested in an attempt to assess the constitutivity of the synthetase enzyme.[25] Cycloheximide inhibited its activity but, surprisingly, when it was added at a low concentration that was insufficient to inhibit protein synthesis appreciably, it actually stimulated this activity. Similar results were noted with amino acid analogues. Light[25] explained this phenomenon on the basis of the ensuing metabolic changes that simulated those in older cultures and encouraged expression of synthetase activity. Indeed the application of growth inhibitory conditions has been used as a general method for the increase of yields of metabolites in industrial fermentations.[26]

The problem concerning the isolation of a purified 6-methylsalicylate synthetase was then attacked with renewed intensity in Lynen's laboratory. This enzyme has now been extensively purified[27,28] after initial stabilization on acetone precipitation, followed by ultracentrifugation and sucrose density gradient centrifugation. It was resolved from fatty acid synthetase at the final stage of purification. It migrated as a homogeneous protein and gave a molecular weight 1.3×10^6 and was similar to fatty acid synthetase in many respects including pH optimum, inhibition by thiol group reagents and its ability to form triacetic acid lactone. The data further indicated that enzyme-bound intermediates were utilized and that two thiol sites were involved (cf. the 'central' and 'peripheral' sites that have already been discussed with respect to fatty acid synthesis in Chapter 2). On these grounds, a reaction sequence was proposed in which the single reduction and dehydration steps occurred at the triacetic acid level. The reactants are all shown in Scheme 6.3 as their oxo derivatives but the intermediates may be held on the enzyme complex with the acetate-derived carbonyl group in the enolic form that renders it non-susceptible to reduction.[29] This feature would explain the retention of this oxygen function despite the presence of a suitable reductase and NADPH.

There is no direct evidence to support the stage at which dehydration occurs but it probably takes place after reduction of the triacetyl residue prior to the next condensation (by comparison with fatty acid synthesis). However, the dehydratase concerned must have a different specificity compared with that normally present in fatty acid synthetase since it yields the 5-oxo-β,γ-*cis*-hexenoyl thioester and is therefore analogous in activity to the β-hydroxydecanoyl-ACP thioester dehydratase involved in the synthesis of *cis*-vaccenate in bacteria.[80]

Other investigators[16,30] have succeeded in isolating cell-free systems capable of synthesizing orsellinic acid (V) and alternariol (IV), phenols derived exclusively from acetate. Orsellinate synthetase activity was isolated from lyophilized (freeze-dried) cells of *P. madriti* and was retained in the high-speed supernatant fraction.[30] An active preparation capable of forming

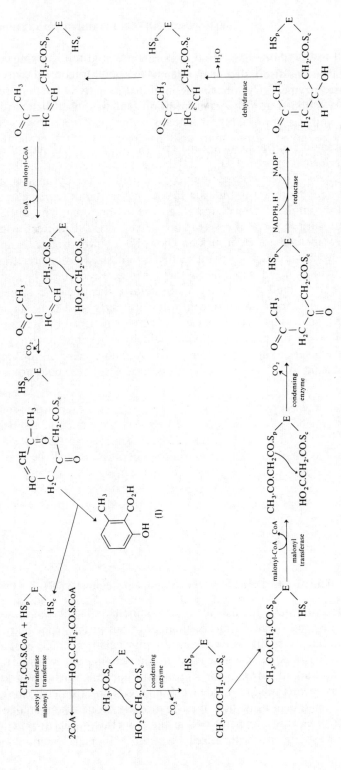

Scheme 6.3 Hypothetical scheme for the synthesis of 6-methylsalicylate on a suitable multienzyme complex

alternariol was prepared from *Alternaria tenuis* after grinding the mycelium with sand. Again, the activity was located in the soluble fraction of the cells.[16]

MODIFICATIONS AND VARIANTS OF ASSEMBLY

Alkylation Reactions

The alternate enolizable methylene groups in the aliphatic intermediate (that is on formation of $-CH=C(OH)-$) are reactive towards alkylating agents such as *S*-adenosylmethionine or dimethylallyl pyrophosphate.[31] There are abundant data to indicate that *C*-methylation occurs before ring closure of the enolic intermediate and release of an aromatic product from the enzyme surface. Several investigators have provided indirect evidence to this effect including McCormick and coworkers[32] who isolated a mutant of a tetracycline-producing organism that synthesized 6-nor-pretetramide (the parent four-ringed naphthacene) and hence tetracyclines lacking the methyl group at the C-6 position[33] (see below). Identification of methyltriacetic acid lactone[20] in culture filtrates also suggested that alkylation could take place at this level. This view was later confirmed by Steward and Packter who showed that 5-methylorcylaldehyde (X) but not orcylaldehyde was converted into gliorosein (XI) by whole cells of *Gliocladium roseum* (Scheme 6.4).[34] The *C*-methyltransferase therefore could not accept

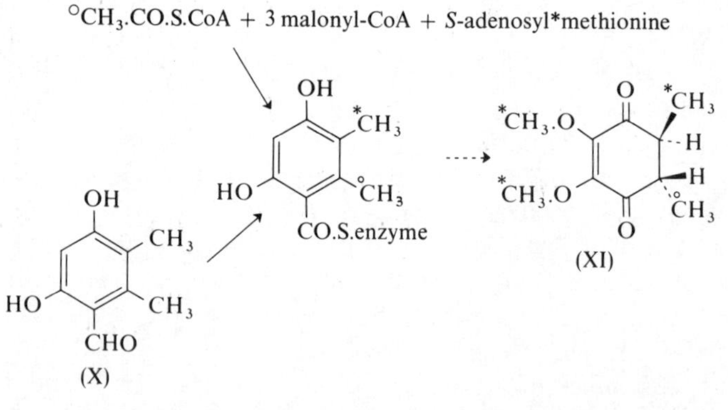

$$^{\circ}CH_3.CO.S.CoA + 3\,malonyl\text{-}CoA + S\text{-}adenosyl*methionine$$

Scheme 6.4   Origin of the methyl groups in gliorosein (XI)

the aromatic orcylaldehyde as substrate. It must be remembered, however, that alkylation may occur at nucleophilic positions in the aromatic ring of many metabolites (usually derived from shikimic acid or related substances) as in the synthesis of the flavonoids and also ubiquinone and vitamin K, etc.

The *O*- and additional *C*-methyl groups of gliorosein arise by transfer from *S*-adenosylmethionine. Evidence from mass spectrometric studies[35] established unambiguously that three hydrogen atoms were retained in each of the three cases (and not two in the *C*-methyl group as previously indicated in experiments with mixed ^{14}C- and ^{3}H-labelled methionine as substrate).[34] The product gave a mass peak corresponding to $M + 9$, where M equals the mass of the molecular ion and is equivalent to the molecular weight. Similarly, Lederer and colleagues[36] demonstrated that all the deuterium atoms from [methyl-$^2H_3$]methionine were recovered in the methyl groups (marked with an asterisk) attached to a phenolic hydroxyl group and aromatic ring of mycophenolic acid (XII), giving rise to a molecular ion of

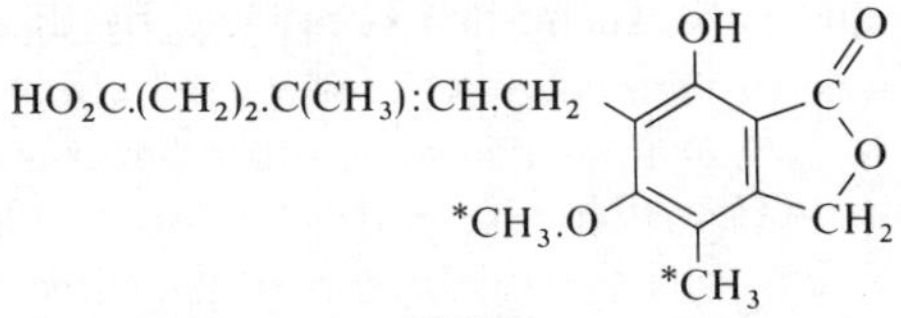

$$HO_2C.(CH_2)_2.C(CH_3):CH.CH_2$$

(XII)

$M + 6$. It is worthy of note that mycophenolic acid contains an additional isoprenoid-derived group attached to a carbon atom in the aromatic skeleton that also originates from the methylene of a malonyl residue.

Removal of Oxygen

Dehydration takes place at the polyketide level by means of a specific dehydratase on an hydroxy acyl intermediate (Scheme 6.3). This aspect will be developed more fully later (Scheme 6.11).

Hydroxylation

The final metabolite secreted into the medium by fungi and streptomycetes often contains hydroxyl groups in positions that are not predicted by the 'acetate hypothesis'. Its nature depends upon the presence of oxygenases in the mycelium, that insert molecular oxygen in *ortho* or *para* orientations to existing hydroxyl substituents, and also O-methyltransferases. The structure of the phenolic acids (I) and (V) is such that they might be the immediate precursors of more complex metabolites. Experiments with whole cells of certain *Aspergilli* sp., for instance, have proved that this is indeed the case, as shown by the conversion of 6-methylsalicylic acid into terreic acid,[37] and orsellinic acid into trihydroxytoluene (XIII)[38] and further

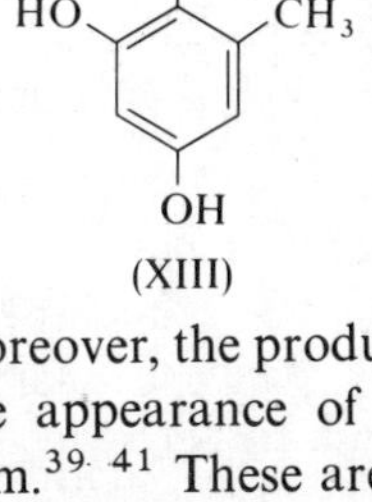

(XIII)

hydroxylated products.[39] Moreover, the production of *p*-diphenols (quinols) is often associated with the appearance of the corresponding quinones after oxidation in the medium.[39, 41] These are usually present in low yield if the medium is rapidly extracted with solvent after removal from the incubator and may generally be considered as artifacts.

Monooxygenases incorporate one atom of oxygen from molecular O_2 per molecule of substrate and release the other as water. A reductant is therefore essential for activity as electron donor and the enzymes are often NAD(P)H-linked. They have been termed 'mixed-function oxygenases' and contain either flavin or pteridine prosthetic groups that mediate in the transfer of hydrogen.[42] They catalyse reactions in which —H or —CO_2H is replaced by an —OH substituent, as exemplified by the conversion of 4-hydroxybenzoate into protocatechuate[43] (equation (2)), and salicylate into catechol[44] by the enzyme salicylate hydroxylase [salicylate, NADH: oxygen oxidoreductase (1-hydroxylating, 1-decarboxylating)] (equation (3)):

$$+ O_2 + NADPH + H^+ \rightarrow \qquad + H_2O + NADP^+ \tag{2}$$

$$+ O_2 + NADH + H^+ \rightarrow \qquad + H_2O + NAD^+ + CO_2 \tag{3}$$

Reduction of the prosthetic group by exogenously supplied cofactor as demonstrated above and the subsequent steps entailing insertion of oxygen are usually tightly coupled in a single bifunctional enzyme.

Tetracyclines

This series of powerful broad-spectrum antibiotics, e.g. oxytetracycline (XIV), that acts against many pathogenic microorganisms is elaborated by species of *Streptomyces*. It furnishes some rich examples of modifications that may arise although full details of their biosynthesis have not been resolved. The initiating carboxamido group attached to C-2 is derived from a malonamoyl unit itself formed from malonyl-CoA by amidation or possibly from acetyl-CoA by carbamoylation. The methyl groups attached to the nitrogen atom and C-6 arise from S-adenosylmethionine while inorganic chloride acts as the source of the chloro-group at C-7 in chlortetracycline,[45] one of the few compounds in Nature which possess a C—Cl bond. In brief, McCormick[46] has proposed that the first stabilized product to emerge from the aromatic synthetase matrix is 6-methylpretetramide (XV), that is, with

(XIV)   (XV)

the C-methyl group incorporated at the polyketide level and the oxygen function at C-8 removed. This then undergoes further modifications in the formation of oxytetracycline (XIV) including hydroxylation at C-4 (and subsequent transamination), C-12α, C-6 and C-5 by means of distinct reactions and transmethylation at the newly formed amino group at C-4.[47] The additional groupings to the primary biogenetic skeleton have been marked in structures (XIV) and (XV). Much of the evidence available for the later stages in tetracycline biosynthesis has been obtained from mutant studies based on accumulation of metabolites and their tentative identification as precursors, followed by their incorporation in wild-type organisms. The precursor activity of these substances has since been confirmed in cell-free systems.

Macrolides

This term describes structures containing the common feature of a large lactone ring.[48] They are produced by many species of *Streptomyces* and occur in Nature as glycosides. They are formed after mycelial growth has ceased[49] and often possess antibiotic activity but, as with many natural products that have been studied, this behaviour may simply reflect the properties of the individual macrolides that have been characterized to date. A particular derivative that is proving of great current interest is rifamycin together with its semi-synthetic variants. These antibiotics whose structure also contain an aromatic ring inhibit bacterial and viral DNA-dependent RNA polymerase at the RNA chain initiation step.[50]

Synthesis of macrolides does not involve a modification of a preformed acetate-derived polyketide but rather relies on a variant on the mode of assembly. Propionyl-CoA acts as primer while its carboxylated product, 2-methylmalonyl-CoA, participates as precursor of the remaining C_3 residues within the branched-chain product. Certain macrolides, however, originate from a mixed propionate/acetate source.[51] This usage of propionate seems restricted to the taxonomically related streptomycetes and mycobacteria but it should be noted that synthesis of their fatty acids is generally accomplished from acetyl-CoA with malonyl-CoA as condensing agent, giving rise to the typical straight-chain acids.[52] Moreover, the C-methyl groups that are present in the branched-chain iso and anteiso acids of these Actinomycetes derive from valine, leucine or isoleucine catabolism while that in 10-methylstearate arises from S-adenosylmethionine (Chapter 3, p. 58). On the other hand, propionyl-CoA is accepted as starter for the provision of odd-numbered fatty acids in all synthetases and as the source of the condensing group in the formation of the starred carbon atoms in the branched moiety of mycocerosic acid (XVI) in *M. tuberculosis*.[53]

$$CH_3.[CH_2]_{18}.CH_2.[*CH(*CH_3).*CH_2]_3.*CH(*CH_3).*CO_2H$$

(XVI)

Investigations into the origin of the C_{21} erythronolide A (XVII), the lactone portion of erythromycin A and C glycosides, indicated that [1-^{14}C, 3-^{3}H]propionate was incorporated into this aglycone with the ^{14}C:^{3}H ratio unchanged.[54] The methyl group was utilized for the synthesis of the C-15 terminal portion and branched methyl groups.[55] Many workers later showed that methylmalonate (presumably acting as its coenzyme A thioester) was specifically converted in *Streptomyces erythreus* into the pre-terminal C_3 units within erythronolide.[56] These results have been interpreted[57] to mean that erythronolide synthesis follows a mechanistically similar process to that of polyacetate-derived compounds (Scheme 6.5). The energy for the process is again derived ultimately from the hydrolysis of ATP in the course of the propionyl-CoA carboxylase reaction which gives rise to methylmalonyl-CoA.[58]

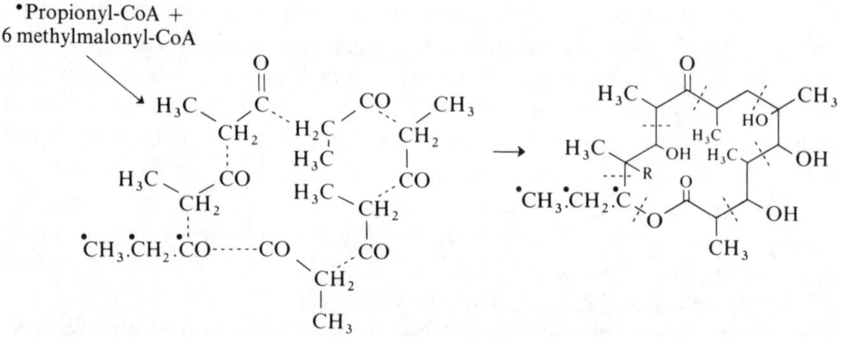

Scheme 6.5   Conversion of propionyl and methylmalonyl thioesters into the terminal and pre-terminal residues of erythronolides. R = H erythronolide B; R = OH erythronolide A (XVII)

FORMATION OF POLYACETYLENES

Polyacetylenes form another characteristic group of polyketides that are synthesized by means of repeated condensation reactions between acyl and malonyl thioesters. They occur in certain higher plants and fungi with C_{18} acetylenic fatty acids fairly prevalent as acyl residues in the seed fats of representatives of the Compositae, Umbelliferae and Araliaceae families.[59] A simple example of a metabolite in this major grouping is ximenynic acid (XVIII) which may contribute over half the total seed fat acids and, as its name implies, contains one double bond ('en') and one triple bond ('yn'). Similarly, the related dienyne (XIX) and enediyne (XX) are also found in

$$CH_3.[CH_2]_4.C\vdots C.CH_2.CH\vdots CH.[CH_2]_7.CO_2H. \quad (XXI)$$

$$CH_3.[CH_2]_5.CH\vdots CH.C\vdots C.[CH_2]_7.CO_2H \quad (XVIII)$$

$$CH_3.[CH_2]_3.[CH\vdots CH]_2.C\vdots C.[CH_2]_7.CO_2H \quad (XIX)$$

$$CH_3.[CH_2]_3.CH\vdots CH.[C\vdots C]_2.[CH_2]_7.CO_2H \quad (XX)$$

association with this acid. Comparison of the formulae in these compounds suggests that the more unsaturated acids may be formed by progressive desaturation giving rise to a series of conjugated triple and double bonds. These confer pronounced and distinctive ultraviolet-absorption spectra on the products which incidentally greatly assist in their identification. Many can be formally derived from linoleic acid ($C_{18:2,\Delta^{9,12}}$) via its acetylenic 12,13-dehydro-derivative, crepenynic acid (XXI),[1] by dehydrogenation and rearrangement. The C_{17} acids probably arise after decarboxylation often followed by modification of the newly formed methyl group.

Several acetylenes with an aromatic function have been isolated from plants and may be formed from the terminal C_7 unit of a polyenoic precursor by dehydration followed by an aldol condensation (Scheme 6.6), in a manner

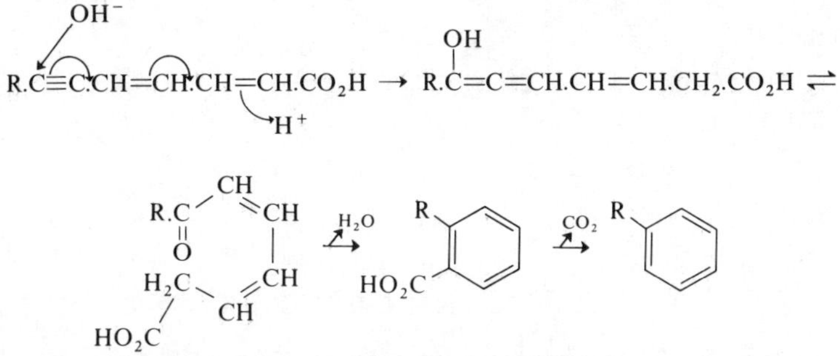

Scheme 6.6   Possible derivation of aromatic acetylenes. R represents an acetylenic side-chain

analogous to the synthesis of phenols.[60] This process would yield products with an o-carboxyl group; 6-alkynylsalicylic acid derivatives have also been identified in Nature.

Polyacetylenes have been isolated from the growth medium of fungi (Basidiomycetes), notably the Order Agaricales, but these tend to have shorter chain-lengths. Odd-numbered products with terminal ethynyl groups are formed after decarboxylation of α,β-acetylenic acids.[61] In general, secretion of polyacetylenes appears greatest when mycelial growth has ceased especially on transfer to non-germinating media or glucose. This pattern is typical of the fungal production of many polyketides.

Two mechanisms have been suggested for the formation of acetylenic bonds. They may be derived from more saturated parent compounds of the linoleoyl-CoA type as evidenced by the conversion of specifically labelled [10-^{14}C]oleate into crepenynic acid (XXI) in the basidiomycete, Tricholoma grammopodium, with retention of label in the C-10 position.[1] The participation of malonyl-CoA as the source of C_2 units had previously been inferred by conversion of diethyl[2-^{14}C]malonate into dehydromatricarianol (XXII)[62] in the same organism and into a related 1,10-dicarboxylic acid.[63] Incorporation into the terminal $CH_3.C\vdots$ fragment of (XXII) was very low; this is derived directly from acetyl-CoA.

$$CH_3.C\vdots C.C\vdots C.C\vdots C.CH:CH.CH_2OH$$

(XXII)

It is possible, however, that certain acetylenes may be formed on a modified synthetase that integrates activity capable of dehydrating enol intermediates or even converting double into triple bonds at specific positions on the chain. Possible examples of these types are illustrated in Scheme 6.7. The former

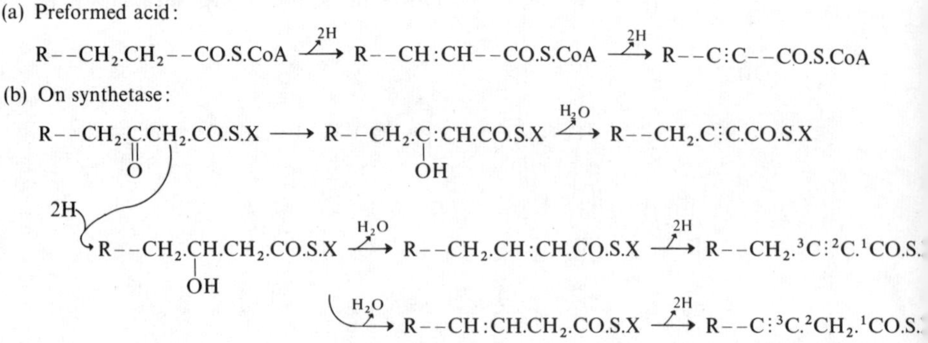

(a) Preformed acid:

$$R--CH_2.CH_2--CO.S.CoA \xrightarrow{2H} R--CH:CH--CO.S.CoA \xrightarrow{2H} R--C\vdots C--CO.S.CoA$$

(b) On synthetase:

$$R--CH_2.\overset{\underset{\|}{O}}{C}.CH_2.CO.S.X \longrightarrow R--CH_2.\overset{\underset{|}{OH}}{C}:CH.CO.S.X \xrightarrow{H_2O} R--CH_2.C\vdots C.CO.S.X$$

$$2H \searrow$$

$$\hookrightarrow R--CH_2.\overset{\underset{|}{OH}}{CH}.CH_2.CO.S.X \xrightarrow{H_2O} R--CH_2.CH:CH.CO.S.X \xrightarrow{2H} R--CH_2.^3C\vdots{}^2C.^1CO.S.$$

$$\hookrightarrow{}_{H_2O} R--CH:CH.CH_2.CO.S.X \xrightarrow{2H} R--C\vdots{}^3C.^2CH_2.^1CO.S.$$

Scheme 6.7   Possible mechanisms for the introduction of triple bonds

reaction denoted as (b) in this scheme has been demonstrated in the reverse direction in extracts from bacteria. Acetylene monocarboxylic (propiolic) acid was hydrolysed to give malonic semialdehyde[64] whereas the 1,2-dicarboxylic acid formed pyruvate after decarboxylation of oxaloacetate generated during the process.[65] These hydration reactions, however, are energetically very favourable and dehydration may be difficult to achieve. On the other hand, the existence of phenolic acetylenes and their probable formation by rearrangement and cyclization reactions adds support for an integrated sequence that utilizes a synthetase complex. Evidence from model reactions, in which concerted decarboxylation of the malonyl residue and elimination of the oxygen function occurs, also tends to confirm this possibility. Moreover, inhibition studies involving addition of cycloheximide or 8-azaguanine to the fungus *Lentinus degener* did not indicate the secretion of unusual acetylenes, under conditions in which the secretion of epoxysuccinate (formed by the action of a monooxygenase enzyme on fumarate) was completely inhibited.[41] The normal enediyne product was present, suggesting that the enzymes or enzyme complex responsible for its synthesis were constitutive and therefore not appreciably affected by the presence of inhibitor. Any change in the pattern of dehydrogenase or dehydratase (or desaturase) activity would have resulted in acetylenic/olefinic products with an altered chromophore and a marked change in their ultraviolet-absorption spectra. The fact that this did not occur may be seen in the dramatic similarity of the spectra obtained from chromatographic fractions derived from the ether-extractable material of the control and inhibited cultures. These are illustrated in Figure 6.1.

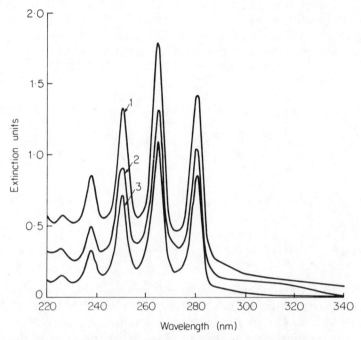

Figure 6.1 Ultraviolet-absorption spectra of the enediyne product isolated from the medium of *Lentinus degener*.
  The spectra were all performed in ethanol and showed identical λ_{max}. The numbers refer to product obtained from control cultures (1), cultures inhibited with cycloheximide (2) and 8-azaguanine (3)

ASSEMBLY PATTERNS

The question has often been asked as to the determining factor for the assembly patterns of the multitudinous aromatic products secreted by fungi. What influences the number of C_2 units engaged, the site of possible reductase activity and eventual release of products with benzene, naphthalene, anthracene, etc., non-linear or heterocyclic ring systems? The condensation of the acetyl and malonyl residues very frequently results in the formation of specific rather than mixtures of products, implying that the intermediates adopt defined configurations during synthesis. Details relating to the properties of the aromatic synthetases still remain open to much speculation but a simplified model may be based on the stereospecific orientations attained by the intermediate enols, controlled by the architecture of the individual synthetase complex. The same pattern would effectively be achieved if the $-CO.CH_2-$ groups were held in a rigid conformation. This model will now be presented, however, in terms of enolic isomers. Let us examine first the simplest case, that of triacetic acid lactone (VII) formation. Assuming that the C-5 oxo group is held as an enolate anion to give the required nucleophilic group for attack on the thioester position (Scheme 6.8), six isomeric forms may be considered: the C-3 oxo group may

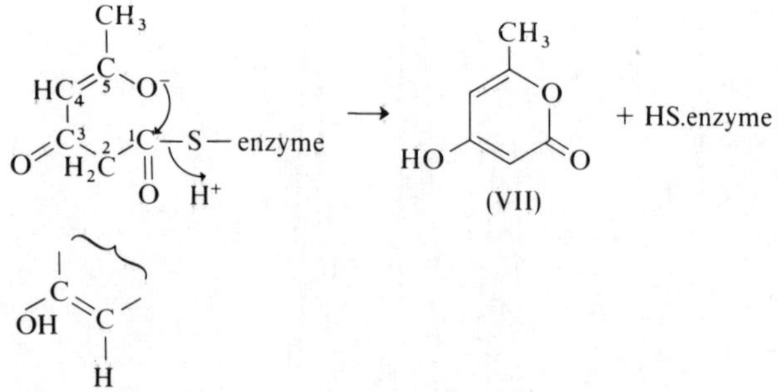

Scheme 6.8   Conversion of the C_6 thioester intermediate into triacetic acid lactone

exist as the carbonyl form or as the *cis*- or *trans*-enolic isomer (with the stereochemistry based on the relative positions of the H— and OH— groups), with the C-5 group fixed as one or other stereoisomer. Lynen has previously mentioned the possibility of enolic intermediates in the course of the condensing enzyme reaction in connection with fatty acid synthesis (see Chapter 2, Scheme 2.2). Examination of space-filling models clearly shows that for synthesis of triacetic acid lactone, the C-5 group must be held in the *trans* form while the C-3 group may be either free to rotate as the carbonyl form or fixed in the *cis* configuration. For the purposes of this argument, the intermediate which possesses H— and OH— groups (or correspondingly carbon substituents) on the same side of the double (or fixed C—C) bond in question is considered the *cis* stereoisomer. Either position of the C-3 group permits the correct geometry for the attack on the thioester carbon atom and subsequent ring closure.

A general development now emerges for the production of more complex metabolites containing a greater number than three C_2 residues. Condensation of the triacetyl group, after transfer to a suitable ('peripheral') binding site, with a further malonyl residue to form a C_8 product is only possible if the former group is stabilized by direction of its C-5 *trans*-enolic oxygen atom away from the thioester region of the molecule during this reaction. The ability to form triacetic acid lactone has only been tested in fatty acid and 6-methylsalicylate synthetases, all of which require NADPH and involve reductive steps. In the presence of this cofactor, therefore, the formation of the hydroxyl group in the vicinity of the C-5 oxygen may aid the process of aromatic synthesis by preventing its reaction with the thioester bond.

Another possibility for the formation of C_8 products, etc., derives from the existence and maintenance of a free oxo group at C-5 of the triacetyl intermediate; this is presented in Scheme 6.9 for the synthesis of tetraacetic

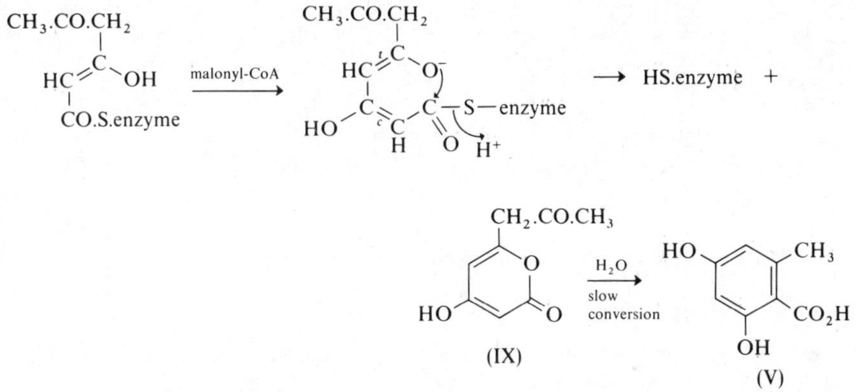

Scheme 6.9   Conversion of the C_8 thioester intermediate into the stabilized tetracetic acid lactone and its hydrolysis into orsellinic acid

acid lactone (IX). This mechanism may account[22] for the presence of small amounts of orsellinic acid (V), formed as a side-product of the synthetase, in the medium of *A. fumigatus* and other fungi. This rate, however, would be insufficient to allow the production of the appreciable amount of orsellinic acid that is found in the medium of a number of *Penicillia* sp., and a more direct process must be operative. This is envisaged in Scheme 6.10 and illustrates a mechanism for the formation of orsellinic acid and javanicin (III), representatives of single and naphthalene ring products.

The photographs of space-filling molecular models depicted in Figure 6.2 demonstrate directly that the appropriate fixture of positions on formation of *trans* and *cis* configurations permits the oxygen function to approach very closely to the reacting hydrogen atom (on carbon atoms that originate from the methylene group of malonyl-CoA) and hence elimination of water and ring closure. The models show the resultant planar configuration of all the carbon atoms on the formation of benzene and naphthalene ring systems.

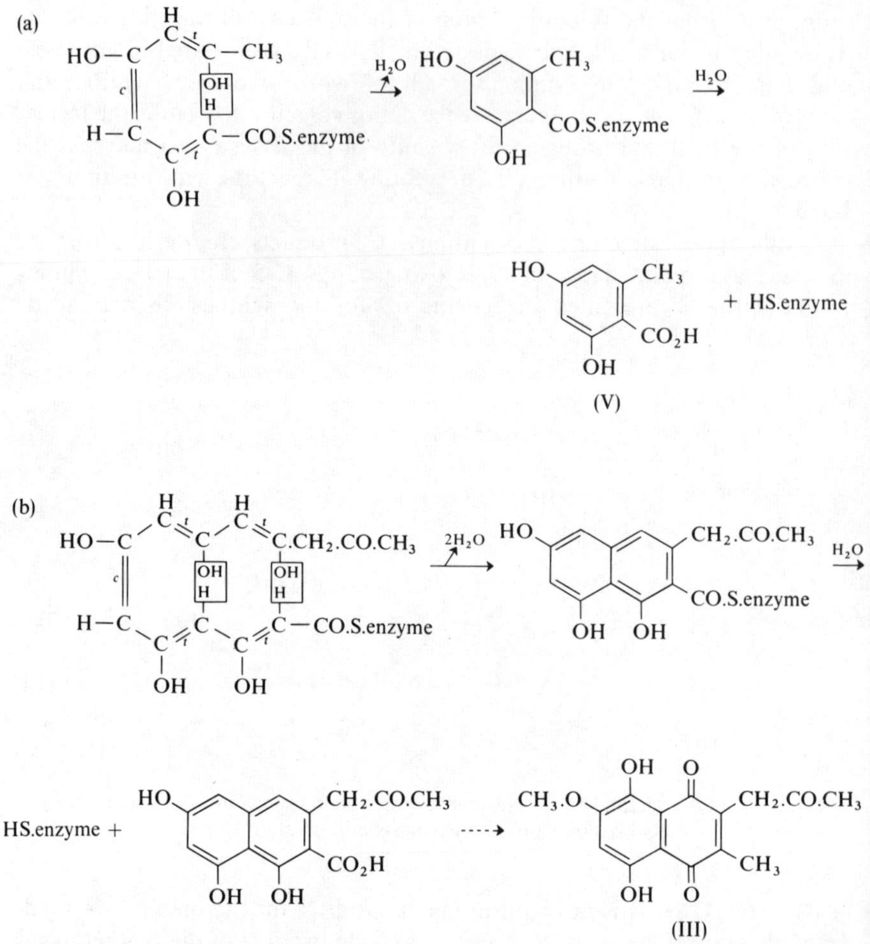

(a)

(b)

(III)

Scheme 6.10   Conversion of the enolic form of (a) the C_8 (*tct*) and (b) the C_{14} (*ttctt*) thioester intermediates into orsellinic acid (V) and javanicin (III) ring systems

The *cis* double bond allows the molecule to attain the shape required for subsequent ring closure. A second *cis* bond would yield a non-linear assembly. A sequence of four *trans* bonds followed by one *cis* bond gives rise to the linear ring product, the 6-methylpretetramide (XV) nucleus of the tetracycline series. Thus the type of ring system achieved is related, in the first instance, to the number of *trans* linkages preceding the formation of a *cis* double bond.

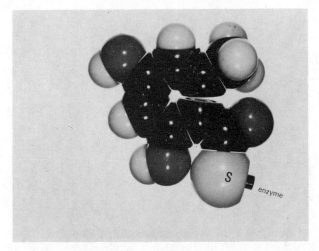

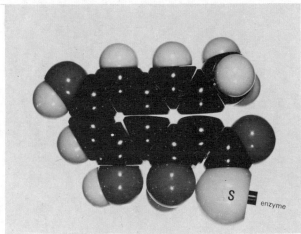

Figure 6.2 Representation of the C_8 and C_{12} thioester
intermediates with the appropriate OH and H groups
removed

Reduction and Dehydration Reactions

The above discussion has ranged over possible mechanisms for various
cyclization patterns but has not taken into account the effect of removal
of oxygen after reduction of a carbonyl group and subsequent dehydration.
Examination of the intermediates given in Scheme 6.10 shows that the
trans-enolic groups related to the terminal methyl or 2-oxopropyl positions
are lost on condensation with the *trans*-enolic hydrogen atoms aligned
near the thioester group. The adjacent oxygen function, however, remains

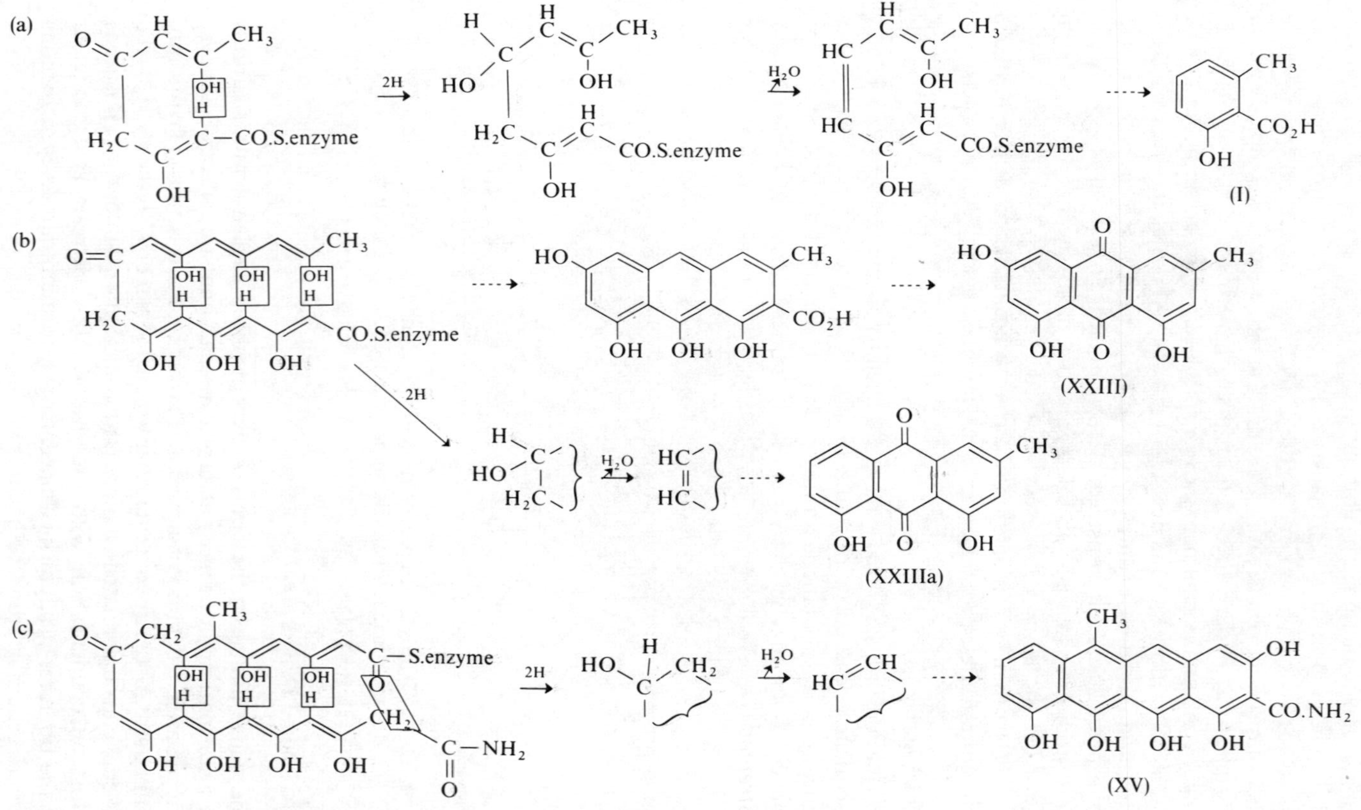

Scheme 6.11   Sites of reductase and dehydratase activity engaged in the removal of oxygen functions in the formation of (a) 6-methyl-salicylic acid (I), (b) desoxy analogue of emodin (XXIIIa) and (c) 6-methylpretetramide (XV)

available for reduction, presumably acting in its capacity as a carbonyl group. (The enol form of β-oxoacyl-ACP substrates does not react with or bind to the reductase enzyme in the *E. coli* fatty acid synthetase.[29]) The reduction product reacts with a dehydratase in the enzyme complex to yield a *cis*-unsaturated derivative, for instance, 6-methylsalicylic acid (I) in *P. patulum*.[27] A similar argument may be followed for the origin of this double bond in the synthesis of the desoxy analogue (XXIIIa) of emodin (XXIII) in the anthraquinone series and also 6-methylpretetramide (XV) (Scheme 6.11). This scheme has been constructed to summarize these effects. Thus *cis*-unsaturation is essential for folding of the molecule and its derivation arises from the activity of a specific β,γ-dehydratase, a component of the aromatic synthetase. Elimination of water from suitable enolic intermediates (if an appropriate enzyme were present) could give acetylenic products.

Ring Closure

Formation of a stabilized ring in the vicinity of the thioester region and consequent release of a phenolic product from the complex may occur by means of a spontaneous aldol condensation reaction[66] or under the influence of an enzyme-derived proton. Evidence from blocked mutants of *Streptomyces* sp., however, favours the latter possibility as protetrone (XXIV) with a 6-demethyl anthracene structure is secreted.[46] The cyclization reaction leading to the normal pretetramide structure would be prevented if the appropriate enzyme were missing.

(XXIV)

REGULATION OF PHENOL SYNTHESIS IN FUNGI

The intriguing question concerning the reason for the fungal secretion of acetate-derived phenols and other products, and the enzymic versatility encountered in their synthesis has often been posed. Indeed, similar questions may be asked about all cells that synthesize 'shunt' or 'secondary' metabolites. They are generally formed late in the growth cycle and seldom have any apparent function in the producing organism. Their secretion, however, may be connected with the maintenance of the environment within the cell as an adjustment to less favourable conditions. Excessive production of phenols by fungi in laboratory cultures is no doubt related to the fact that these are generally grown on media rich in glucose (5 per cent in the case of

Raulin-Thom or Czapek-Dox media) but may also reflect a lack of control on its entry into the mycelium and subsequent metabolism *via* glycolysis, etc. Regulation of transport into the cells does not seem to play an important role in economy of fungal metabolism. Moreover, the initial enzymes that dictate the rate of entry of glucose into metabolic pathways do not appear to be susceptible to feed-back control mechanisms and are available for some considerable time after depletion of an essential nutrient.

A rationale that would explain phenol production is that deprivation of this nutrient (or metal ion) might encourage a yield of acetyl-CoA and tricarboxylic acid cycle intermediates in excess of that required for growth. Under these circumstances, a continued flux through glycolysis with production of acetyl-CoA and a consequential increase in the rate of malonyl-CoA formation might enhance fatty acid (and lipid) and also phenol synthesis. With *Gibberella fujikuroi*, for instance, exhaustion of nitrogen from the medium in the presence of excess glucose causes cell proliferation to cease and triglycerides to accumulate in the mycelium.[67] In many fungi the response to this situation of increased substrate availability also promotes the secretion of acetate- (or mevalonate) derived products such as the anthraquinone pigments in *P. islandicum*.[68] On the other hand, maintenance of a supply of fresh medium and hence balanced growth may prevent the occurrence of these metabolic events.

In the natural environment such abnormally high glucose concentrations are presumably not encountered. The capacity to produce these metabolites under normal conditions in the soil (and this is still a matter for conjecture) may be stimulated by internal factors such as the build-up of primary intermediates after balanced growth has ceased.[31] The formation of phenols (and fatty acids, of course) is a means by which the internal milieu of the cell would be restored, for example, by releasing coenzyme A in the course of the acetyl-CoA plus malonyl-CoA condensation reactions. This phenomenon may then be considered analogous to fatty acid synthesis and resultant lipogenesis by normal animals on eating excess carbohydrate, or ketogenesis in the liver of starving or diabetic animals. In this latter case, two molecules of acetyl-CoA are effectively converted into acetoacetate. Thus as Bu'Lock first proposed,[31] any selective advantage gained by the organism appears related to its ability to adjust to an unfavourable environment such as a deficiency in the supply of N or P rather than the physiological activity of the products in the extra-cellular phase on the surrounding heterogeneous microflora. A further significant feature may lie in their removal from the mycelium, preventing feed-back inhibition on the enzymes concerned with their synthesis and ensuring a continuous flow of the terminal product. If these processes do occur in the wild state, the considerable intrinsic energy of the phenolic metabolites would not be lost to Nature but could be utilized by bacteria after ring cleavage and degradation.

Similarly, the wood-rotting fungi may decompose plant phenolics such as flavonoids, tannins and lignins.[69]

There is, however, some evidence that metabolites accumulated within the mycelium may exert a protective function. Certain fungi, including *Aspergilli* sp., are resistant to microbial degradation in their natural environment because their cell-walls are not readily susceptible to enzymic digestion by glucanases and chitinase.[70] This phenomenon is probably related to the presence of melanin-type material in the hyphal and conidial walls since melanin-less mutants are not resistant to hydrolysis.[71] Thus, the ability to produce these phenolic and quinonoid pigments (and possibly also similar substances formed from acetate) may confer some protection on the producing organism and increase its ability to survive. In addition, acetate-derived polyacetylenes are found in the sporophores of fruiting bodies from several basidiomycetes in the wild and occur naturally, of course, in higher plants. A further interesting observation in this context was made by Anchel who noted that a polyacetylene isolated from a basidiomycete which is naturally associated with the mycorrhiza of pine proved antagonistic to a known pathogen of this species.[72]

Constitutive and Induced Enzymes

The genetic capability leading to the formation of the enzymes concerned is obviously a prerequisite for the synthesis of any product. It is only recently that a distinction has been made between the involvement of constitutive and induced enzymes in the metabolic developments leading to the formation of phenols, with the use of inhibitors of protein synthesis. In general, the results indicate that primary intermediary metabolites, in this instance acetyl-CoA and malonyl-CoA, may be converted into an initial key product of an aromatic nature by enzymes already present in the mycelium. This in turn often acts as precursor for a range of further metabolites, probably as a result of induction of 'new' enzymes, usually of the oxygenase type.

Bu'Lock and coworkers[73] demonstrated that addition of these inhibitors to the medium of *P. uritcae* (*P. patulum*) exerted different effects dependent on the time of supplementation. When added during replicatory growth or at the onset of 6-methylsalicylate production, synthesis of further metabolites such as gentisaldehyde (XXV) and patulin (XXVI) (and presumably of the enzymes that gave rise to them) was inhibited. Synthesis of 6-methylsalicylate itself was not affected and some *p*-toluquinol (XXVII) was also present, formed *via* the intermediacy of *m*-cresol (Scheme 6.12), even when the inhibitors were administered before the onset of phenol synthesis. However, 6-methylsalicylate was not converted into other products when added to young cultures. Thus despite the apparent presence of the synthetase during the growth phase and its presumed constitutivity, its activity was not expressed until a later phase when different nutritional conditions

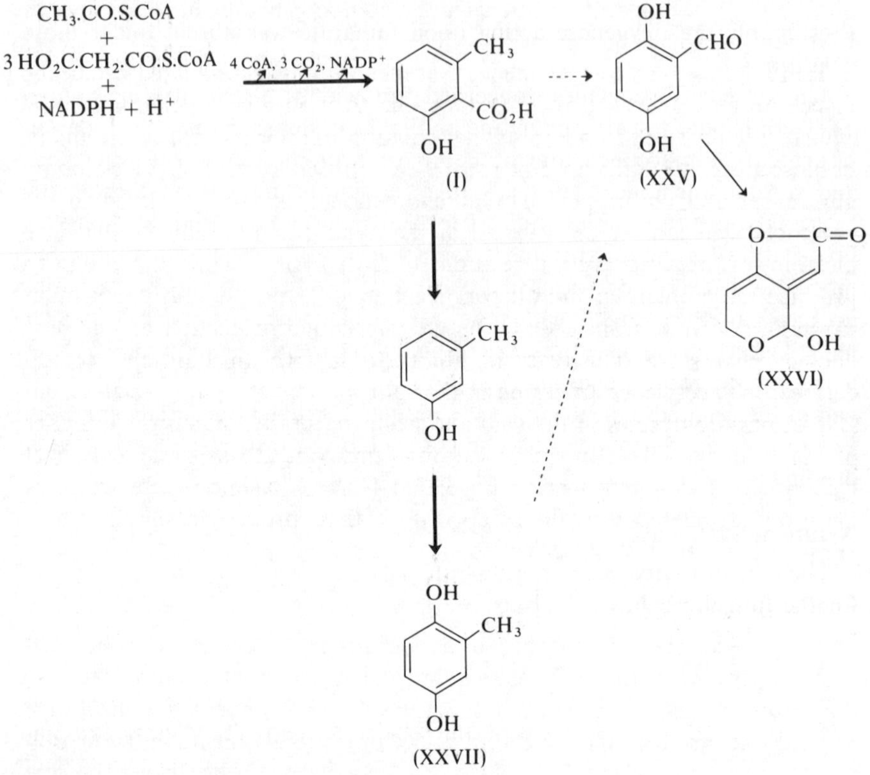

CH3.CO.S.CoA
+
3 HO2C.CH2.CO.S.CoA
+
NADPH + H+
4 CoA, 3 CO2, NADP+
CH3
CO2H
OH
(I)
OH
CHO
OH
(XXV)
O
C=O
O
OH
(XXVI)
CH3
OH
OH
CH3
OH
(XXVII)

Scheme 6.12 Formation of metabolites from acetyl-CoA and malonyl-CoA. Reactions drawn with heavy arrows indicate those remaining after treatment with inhibitors of protein synthesis

applied. Similar results were obtained on treatment of *A. fumigatus* with cycloheximide.[74] Synthesis of trihydroxytoluene (XIII) was not affected but tetrahydroxytoluene and fumigatol (an *O*-methyl derivative) secretion were minimal. The implication therefore is that only monooxygenase enzymes and probably those that act later in the metabolic sequence are inducible in this organism and *P. urticae*. Surprisingly, however, the expected product of the constitutive synthetase in *A. fumigatus*, orsellinic acid (V), did not accumulate in amounts greater than that in the controls; possibly the particular oxygenase responsible for the formation of trihydroxytoluene is integrated within the complex. A selective and marked effect of cycloheximide and 8-azaguanine on the synthesis of epoxysuccinate by *L. degener* was also noted but supplementation of the medium with these inhibitors at the same stage of growth did not affect synthesis of the acetate-derived polyacetylene.[41]

Presumably the oxygenase acting upon fumarate was absent under those conditions.[75]

The oxygenase enzymes concerned are widely distributed in Nature. They contribute towards many metabolic processes, including degradation of aromatic amino acids and transformation of cholesterol into bile acids and steroid hormones. They also play an important role in the hepatic hydroxylation and conjugation of foreign compounds of an aromatic or lipid nature *via* the microsomal cytochrome P-450 system into soluble and more readily excretable forms (detoxication);[76,77] this explains the tolerance acquired after treatment with many drugs. Thus the insertion of further hydroxyl groups (and expression of methyltransferase activity) appears to be a general response to the provision of preformed aromatic products within the cell and depends upon the induction of the appropriate enzymes. The term 'secondary' metabolite might usefully be restricted to this type of product.

Nature of Synthetase

The constitutivity of aromatic synthetases leaves unexplained the reason for the fungal synthesis of phenols whether this is in addition or an alternative to fatty acid formation. Recent experiments in the Author's Laboratory[78] have shown that even on depletion of nitrogen and other nutrients, by transfer of washed mycelium to 5 per cent glucose solution, *A. fumigatus* does not synthesize phenols for some considerable time. Therefore, metabolic events that ensue later must be responsible for the initiation of phenol production, presumably by some modification of the fatty acid synthetase or the nature of its assembly. This could be envisaged for the synthetases that yield orsellinic acid (V), trihydroxytoluene (XIII) and alternariol (IV), which do not require reductase activity, as resulting from inhibition of this component in the existing synthetase enzyme or, more probably for these and other enzymes, by elaboration of a modified multienzyme complex. Prevention of reductase activity in fatty acid synthetases and indeed in 6-methylsalicylate synthetase by omission of NADPH simply gives rise to triacetic acid lactone (VII); the facility enabling the triacetyl residue to react with further malonyl groups does not appear to be present. This would be the case if the enolate anion were held in the 'wrong' configuration on the enzyme surface. Examination of the enzymological and physico-chemical properties of the fatty acid and aromatic synthetases isolated from a given organism should present valuable information regarding the relationship between the functional activities of the two enzymes and provide a deeper understanding of these processes.

It has been established that 6-methylsalicylate synthetase is a separate protein from fatty acid synthetase in *P. patulum*, with a considerably smaller molecular weight.[27] Since β-oxoacyl reductase activity is present in both enzymes, it may be postulated that control of formation of 6-methylsalicylate synthetase cannot be exerted at this point but rather at the level of elaboration from the enzymes normally utilized for fatty acid synthesis, although it is extremely unlikely that a pool of these enzymes would be free in the cytoplasm. If the same enzymes are involved, it seems that under these circumstances some exert rather different activities, for example, the condensing enzyme is capable of transferring acetoacetyl and other non-fully reduced acyl residues to the cysteine binding site, with additional capacity to form enolic products. Enoyl reductase which requires a *trans*-α,β-unsaturated substrate may be a component but this enzyme is not presented with a suitable substrate. Although both synthetases in *P. patulum* (*P. urticae*) can form triacetic acid lactone, the aromatic synthetase is considerably more active in this respect,[27] possibly reflecting a greater ability for its condensing enzyme to transfer the non-reduced acetoacetyl derivative.

The major notable difference between the enzyme 'pairs', however, lies in the presence of a *cis*-β,γ-dehydratase in the aromatic synthetase in *P. patulum* and other fungi, probably a monooxygenase (that gives rise to trihydroxytoluene as product) in *A. fumigatus* and a *C*-methyltransferase in the enzymes concerned with the formation of gliorosein (XI) and tetracyclines. Studies with inhibitors of protein synthesis demonstrated that the activity of these 'additional' enzymes remain expressed even when *de novo* synthesis cannot take place.[73,74] However, it is equally apparent that the aromatic synthetase containing these enzymes must be related in some way to fatty acid synthetase (or conceivably its component enzymes) although the latter does not possess the ability to generate phenols. The extra enzyme must therefore be inactive during the early stages of mycelial growth.

Fatty acid synthetase from yeast has a very high molecular weight compared with that in animal tissues and appears to be trimer.[79] Fatty acid synthetase from *P. patulum* also has a high molecular weight ($2·6 \times 10^6$) and is exactly twice the value for the molecular weight of 6-methylsalicylate synthetase ($1·3 \times 10^6$) present in the same organism.[28] Possibly then, the β,γ-dehydratase is already present as a component of the fatty acid synthetase, which we might consider for the sake of argument as the dimeric form, but its activity is not expressed in this complex. Initiation of phenol synthesis, that is activition of this enzyme, might then be caused by resolution into the monomeric units after binding to a regulatory metabolite that is available above a threshold concentration (see Chapter 1, Figure 1.1) at this stage of the growth phase. The active site of the dehydratase which had previously been masked or inaccessible may now become exposed as a result of changes in the conformation of the protein. It is very interesting to note at this point that a β,γ-dehydratase acting specifically at the C_{10} level is intimately associated with bacterial fatty acid synthetases and is responsible for the insertion of the single double bond during the synthesis of *cis*-vaccenic acid ($C_{18:1}$).[80] A similar argument may be applied to the sudden expression of synthetase activity leading to the formation of trihydroxytoluene in *A. fumigatus*, the primary *C*-methylated aromatic product in the synthesis of gliorosein and tetracyclines, and other substances. With regard to orsellinate or alternariol synthetases, their activity may be influenced by disaggregation of the fatty acid synthetase in the fungi concerned but there is no need in these instances to postulate the presence of a new enzyme. Disaggregation would apparently be accompanied by loss of β-oxoacyl thioester reductase capacity (possibly arising from conversion of the substrates into enolic isomers) but the activity of the acetyl and malonyl transferases and condensing enzyme would not seem to be affected appreciably. A further point to consider is that provided disaggregation does not proceed to completion, fatty acid synthetase activity would still be expressed. Indeed, the latter process has been tested for directly in *A. fumigatus* where fatty acid synthetase activity remains for a considerable time after nitrogen depletion of the medium (brought about by transfer of the mycelium to glucose solution) although its specific activity is much decreased.[78]

Simple analogies already exist where the enzymic behaviour of certain proteins is greatly modified after binding non-covalently with each other. Mutual conformational changes are induced on combination, for instance, of the sub-units of tryptophan synthetase (EC 4.2.1.20) or the components of lactose synthetase. In the former case which is concerned with the terminal reaction responsible for tryptophan synthesis, the active enzyme consists of two proteins which individually are only capable of trace activity in catalysing the successive half-reactions involved (formation and utilization of the indole intermediate). Maximum activity for either of these reactions and for the overall conversion of indoleglycerol 3-phosphate plus serine into tryptophan, with the release of glyceraldehyde 3-phosphate, is only achieved by the active two-component complex.[81] Similarly using lactose synthetase from lactating mammary gland as an example, promotion of the ability to synthesize lactose is found on aggregation of the two proteins, the membrane-bound *N*-acetyllactosamine synthetase (whose activity is poorly exhibited on aggregation) and α-lactalbumin, a milk whey protein.[82] In this situation, the acceptor specificity of the enzyme for the UDP-galactose substrate changes from *N*-acetylglucosamine to glucose in the presence of α-lactalbumin.

A highly simplified diagram representing the proposed model for the synthesis of acetate-derived phenols in fungi appears in Scheme 6.13. The hatched areas denote any additional component that may be present; the open areas correspond to the enzymes present in fatty acid synthetase which have been described previously in Chapter 2. Finally, a possible sequence of events leading to the production of these compounds, and reflecting the intracellular changes that may occur, is briefly expressed in Scheme 6.14.

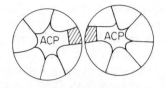 Effector metabolite

Active fatty acid synthetase in   Inactive fatty acid synthetase;
phenol-secreting fungi;

No aromatic synthetic activity   Aromatic synthetic activity

Scheme 6.13 Hypothetical relationship between fatty acid and aromatic synthetases in fungi. ACP; acyl carrier protein

Depletion of nitrogen or other vital nutrient from the medium; glucose remains

Cessation of growth; rate of acetyl-CoA formation is maintained or increased; metabolic changes build up in mycelium and lead to an increased (or decreased) concentration of an effector metabolite

The metabolite binds to fatty acid synthetase and causes disaggregation

Monomeric units formed that possess aromatic synthetic activity and often exhibit an 'additional' enzyme activity over that expected for the components of fatty acid synthetase; formation of primary phenolic product; secretion into the medium

Frequent induction of monooxygenase and other enzymes that gives rise to insertion of extra hydroxyl groups (that are often O-methylated); secretion of metabolites into the medium.

Scheme 6.14 Working hypothesis to demonstrate the possible sequence of events leading to the secretion of acetate-derived products by fungi

REFERENCES

1. Bu'Lock, J. D., and Smith, G. N., *J. Chem. Soc. C*, 332 (1967)
2. Collie, J. W., *J. Chem. Soc.*, **91**, 787, 1806 (1907)
3. Robinson, R. In *The Structural Relations of Natural Products*. Clarendon Press, Oxford, 1955
4. Birch, A. J., and Donovan, F. W., *Aust. J. Chem.*, **6**, 360 (1953)
5. Birch, A. J., Massy-Westropp, R. A., and Moye, C. J., *Aust. J. Chem.*, **8**, 539 (1955)
6. Birch, A. J., Massy-Westropp, R. A., Rickards, R. W., and Smith, H., *J. Chem. Soc.*, 360 (1958)
7. Gatenbeck, S., and Bentley, R., *Biochem. J.*, **94**, 478 (1965)
8. Thomas, R. *Biochem. J.*, **78**, 748 (1961)
9. Gatenbeck, S., and Mosbach, K., *Acta Chem. Scand.*, **13**, 1561 (1959)
10. Watkin, J. E., Underhill, E. W., and Neish, A. C., *Can. J.Biochem. Physiol.*, **35**, 229 (1957)

11. Lynen, F., *J. Cell. Comp. Physiol.*, **54,** Suppl. 1, 33 (1959)
12. Lynen, F., and Tada, M., *Angew. Chem.*, **73,** 513 (1961)
13. Bu'Lock, J. D., Smalley, H. M., and Smith, G. N., *J. Biol. Chem.*, **237,** 1778 (1962)
14. Light, R. J., *J. Biol. Chem.*, **242,** 1880 (1967)
15. Mosbach, K., *Naturwissenschaften*, **48,** 525 (1961)
16. Gatenbeck, S., and Hermodsson, S., *Acta Chem. Scand.*, **19,** 65 (1965)
17. Gatenbeck, S., *Biochem. Biophys. Res. Commun.*, **6,** 422 (1961)
18. Douglas, J. L., and Money, T., *Tetrahedron*, **23,** 3545 (1967)
19. Harris, T. M., Harris, C. M., and Light, R. J., *Biochim. Biophys. Acta*, **121,** 420 (1966)
20. Acker, T. E., Brenneisen, P. E., and Tannenbaum, S. W., *J. Amer. Chem. Soc.*, **88,** 834 (1966)
21. Yalpani, M., Willecke, K., and Lynen, F., *Eur. J. Biochem.*, **8,** 495 (1969)
22. Bentley, R., and Zwitkowits, P. M., *J. Amer. Chem. Soc.*, **89,** 681 (1967)
23. Lynen, F., *Fed. Proc.*, **20,** 941 (1961)
24. Light, R. J., and Hager, L. P., *Arch. Biochem. Biophys.*, **125,** 326 (1968)
25. Light, R. J., *Arch. Biochem. Biophys.*, **122,** 494 (1967)
26. Vining, L. C., and Taber, W. A. In *Biochemistry of Industrial Microorganisms*, p. 341. Ed. by Rainbow, C., and Rose, A. H., Academic Press Inc., London and New York, 1963.
27. Dimroth, P., Walter, H., and Lynen, F., *Eur. J. Biochem.*, **13,** 98 (1970)
28. Dimroth, P., Greull, G., Seyffert, R., and Lynen, F., *Hoppe-Seyler's Z. Physiol. Chem.*, **353,** 126 (1972).
29. Schulz, H., and Wakil, S. J., *J. Biol. Chem.*, **246,** 1895 (1971)
30. Gaucher, G. M., and Shepherd, M. G., *Biochem. Biophys. Res. Commun.*, **32,** 664 (1968)
31. Bu'Lock, J. D. In *Biosynthesis of Natural Products*, p. 9. McGraw-Hill Book Co., London, 1965
32. McCormick, J. R. D., Sjölander, N. O., Hirsch, U., Jensen, E. R., and Doerschuk, A. P., *J. Amer. Chem. Soc.*, **79,** 4561 (1957)
33. McCormick, J. R. D., Johnson, S., and Sjölander, N. O., *J. Amer. Chem. Soc.*, **85,** 1692 (1963)
34. Steward, M. W., and Packter, N. M., *Biochem. J.*, **109,** 1 (1968)
35. Lenfant, M., Farrugia, G., and Lederer, E., *C.R. Acad. Sci. Ser. C*, **268,** 1986 (1969)
36. Jauréguiberry, G., Farrugia-Fougerouse, G., Audier, H., and Lederer, E., *C.R. Acad. Sci. Ser. C.*, **259,** 3108 (1964)
37. Read, G., Westlake, D. W. S., and Vining, L. C., *Can. J. Biochem.*, **47,** 1071 (1969)
38. Packter, N. M., *5th IUPAC International Symposium of Natural Products*, C38, p. 126, 1968
39. Packter, N. M., *Biochem. J.*, **98,** 353 (1966)
40. Packter, N. M., and Steward, M. W., *Biochem. J.*, **102,** 122 (1967)
41. Packter, N. M., *Biochem. J.*, **114,** 369 (1969)
42. Hayaishi, O. *Annu. Rev. Biochem.*, **38,** 21 (1969)
43. Hesp, B., Calvin, M., and Hosokawa, K., *J. Biol. Chem.*, **244,** 5644 (1969)
44. Katagiri, M., Takemori, S., Suzuki, K., and Yasuda, H., *J. Biol. Chem.*, **241,** 5675 (1966)
45. Jarai, M., Jozsa, G., and Kollar, J., *Nature (London)*, **204,** 1307 (1964)
46. McCormick, J. R. D., In *Biogenesis of Antibiotic Substances*, p. 73, Ed. by Vaněk, Z., and Hošťálek, Z., Academic Press Inc., London, 1965

47. McCormick, J. R. D., Joachim, U. H., Jensen, E. R., Johnson, S., and Sjölander, N. O., *J. Amer. Chem. Soc.*, **87**, 1793 (1965)
48. Woodward, R. B., *Angew. Chem.*, **69**, 50 (1957)
49. Bu'Lock, J. D., *Advan. Applied Microbiol.*, **3**, 293 (1961)
50. Wehrli, W., and Staehelin, M., *Bacteriol. Rev.*, **35**, 290 (1971)
51. Birch, A. J., Djerassi, C., Dutcher, J. D., Majer, J., Perlman, D., Pride, E., Rickards, R. W., and Thompson, P. J., *J. Chem. Soc.* 5274 (1964)
52. Brindley, D. N., Matsumura, S., and Bloch, K., *Nature (London)*, **224**, 666 (1969)
53. Gastambide-Odier, M., Delauměny, J. M., and Lederer, E., *Biochim. Biophys. Acta* **70**, 670 (1963)
54. Grisebach, H., Achenbach, H., and Grisebach, U. C., *Naturwissenschaften*, **47**, 206 (1960)
55. Kaneda, T., Butte, J. C., Taubman, S. B., and Corcoran, J. W., *J. Biol. Chem.*, **237**, 322 (1962)
56. Grisebach, H., Hofheinz, W., and Achenbach, H., *Z. Naturf.*, **17b**, 64 (1962); Friedman, S. M., Kaneda, T., and Corcoran, J. W., *J. Biol. Chem.*, **239**, 2386 (1964)
57. Corcoran, J. W. In *Biogenesis of Antibiotic Substances*, p. 131. Ed. by Vaněk, Z., and Hošťálek, Z., Academic Press Inc., London, 1965
58. Wawszkiewicz, E. J., and Lynen, F., *Biochem. Z.*, **340**, 213 (1964)
59. Jones, E. R. H., *Proc. Chem. Soc.*, 199 (1960)
60. Bu'Lock, J. D., *Progr. Org. Chem.*, **6**, 117 (1964)
61. Gardner, J. N., Lowe, G., and Read, G., *J. Chem. Soc.*, 1532 (1961)
62. Bu'Lock, J. D., and Smalley, H. M., *J. Chem. Soc.*, 4662 (1962)
63. Jones, E. R. H., Lowe, G., and Shannon, P. V. R., *J. Chem. Soc. C*, 144 (1966)
64. Yamada, E. W., and Jakoby, W. B., *J. Biol. Chem.*, **234**, 941 (1959)
65. Yamada, E. W., and Jakoby, W. B., *J. Biol. Chem.*, **233**, 706 (1958)
66. Light, R. J., Harris, T. M., and Harris, C. M., *Biochemistry*, **5**, 4037 (1966)
67. Borrow, A., Jefferys, E. G., Kessel, R. H. J., Lloyd, E. C., Lloyd, P. B., and Nixon, I. S., *Can. J. Microbiol.*, **7**, 227 (1961)
68. Gatenbeck, S., and Sjöland, S., *Biochim. Biophys. Acta*, **93**, 246 (1964)
69. Dagley, S., *Sci. Progr.*, **53**, 381 (1965)
70. Bloomfield, B. J., and Alexander, M., *J. Bacteriol.*, **93**, 1276 (1967)
71. Kuo, M. J., and Alexander, M., *J. Bacteriol.*, **94**, 624 (1967)
72. Anchel, M. In *Antibiotics*, Vol. II, p. 189. Ed. by Gottlieb, D., and Shaw, P. D., Springer-Verlag, Berlin, Heidelberg and New York, 1967
73. Bu'Lock, J. D., Shepherd, D., and Winstanley, D. J., *Can. J. Microbiol.*, **15**, 279 (1969)
74. Packter, N. M. *6th Meet. Fed. Eur. Biochem. Soc. Madrid*, Abstr. 914, p. 283 (1969)
75. Wilkoff, L. J., and Martin, W. R., *J. Biol. Chem.*, **238**, 843 (1963)
76. Talalay, P., *Annu. Rev. Biochem.*, **34**, 347 (1965)
77. Conney, A. H., *Pharmacol. Rev.*, **19**, 317 (1967)
78. Packter, N. M., and Ward, A. C., *Biochem. J.*, **127**, 14P (1972)
79. Lynen, F., *Biochem. J.*, **102**, 381 (1967)
80. Rando, R. R., and Bloch, K., *J. Biol. Chem.*, **243**, 5627 (1968)
81. Reed, L. J., and Cox, D. J., *Annu. Rev. Biochem.*, **35**, 57 (1966)
82. Brodbeck, U., Denton, W. L., Tanahashi, N., and Ebner, E. E., *J. Biol. Chem.*, **242**, 1391 (1967)

CHAPTER 7

Biosynthesis of Isoprenoid-derived Compounds Formed via *Farnesyl Pyrophosphate: the Sterols*

INTRODUCTION

A great variety of isoprenoid-derived compounds are widely distributed in Nature and include several classes of metabolites of diverse chemical structure: sterols and the various substances formed from them, carotenoids and vitamin A, rubber and polyprenols, and the hydrocarbon portions of such compounds as chlorophyll, ubiquinone, plastoquinone, vitamins E and K. The sterol skeleton is formed from squalene which is synthesized by a condensation reaction between two molecules of farnesyl pyrophosphate. This chapter will be devoted to a discussion of its mode of synthesis while formation of other isoprenoid-derived compounds from the C_{20} geranyl-geranyl pyrophosphate and higher homologues will be covered in the next chapter.

Cholesterol (I) is an alicyclic compound whose structure includes four fused rings with a single hydroxyl group at C-3 and unsaturation at C-5. It is a white solid with high melting point and the physical characteristics of lipids.

(I)

However, it may be readily resolved from the saponifiable components of a tissue or lipid extract after alkaline hydrolysis followed by extraction with organic solvents. The remaining lipids are converted into water-soluble products by this procedure. Cholesterol is the major sterol in the tissues of vertebrates and is especially abundant in brain and nervous tissue where it is associated in myelin with phospholipid as micelles (aggregates in which the polar ends of the constituent molecules are directed outwards). It is also found in other membranous structures but is present at a low concentration in mitochondrial lipids.

Cholesterol and its esters (formed by transfer of an acyl residue from acyl-CoA or lecithin) are carried through the circulatory system combined with lipoproteins; the free sterol may occasionally be deposited, often in nearly pure form, in gall-stones. Cholesterol has tremendous physiological importance in the tissues of higher animals as the principal precursor of many series of essential metabolites, namely the bile acids, adrenocortical hormones and the sex hormones synthesized in the ovary and testis. It is often accompanied by small amounts of precursors including 7-dehydrocholesterol (provitamin D_3) and catabolic products. An appreciation of the pathway by which cholesterol is synthesized (which has often been gained from studies with liver preparations) is therefore essential for a proper understanding of the means by which steroid-derived products are formed in the appropriate tissues.

Other sterols are also found elsewhere in Nature and possess the same hydrocarbon skeleton as cholesterol. The predominant sterol in yeasts and fungi is ergosterol (II) which possesses additional double bonds at C-7,

(II)

and hence a characteristic ultraviolet-absorption spectrum, and C-22 in the side-chain. An extra C-methyl group is also present (C-28). The phytosterols are limited to plant tissues and algae and resemble ergosterol in that additional alkyl groups are attached to the side-chain. Sterols are not found in bacteria.

Historical Review

Bloch and Rittenberg[1,2] established in their early classical experiments that both ^{2}H- and ^{13}C-labelled acetate (containing non-radioactive isotopes) were incorporated into cholesterol (I) in rats and mice. Later work[3] confirmed that [^{14}C]acetate was incorporated into this product in liver slices. Degradative studies on the radioactive material indicated that the carbon atoms in the side-chain were arranged in a definite pattern and that each atom was formed from either the methyl (m) or carboxyl (c) carbon atom of acetate:

Subsequently it was found that [^{14}C]acetate was also converted into squalene, a C_{30} hydrocarbon, both *in vivo*[4] and in preparations derived from rat liver and hen ovary.[5] This product in turn was effectively incorporated into cholesterol after feeding to mice.[6] The outstanding studies of Cornforth and Popják[7] confirmed that the labelling pattern was as expected and again the origin of every carbon atom was assigned to acetate. Next, they extended their techniques and devised a carbon by carbon degradation of cholesterol, isolated from liver slices that had been incubated with [1-^{14}C]- and [2-^{14}C] acetate. After examining their origin, they concluded that each individual carbon atom arose from acetate. A repeating C_5 isoprenoid component appeared to be involved as the major condensing unit in its biosynthesis. The [^{14}C]cholesterol contained 12 equally labelled carbon atoms from [1-^{14}C]acetate (c) and 15 from [2-^{14}C]acetate (m), as shown in (Ia):

(Ia)

Thus it was established that acetate was the distal precursor of squalene, cholesterol and presumably other steroids. Moreover, the initial cyclization product of squalene was characterized as lanosterol, a 3-hydroxy-C_{30} sterol, indicating that the probable sequence of reactions might be acetate (C_2) → isoprenoid unit (C_5) → squalene (C_{30}) → lanosterol (C_{30}) → cholesterol (C_{27}).[8,9] However, these results did not afford any information concerning the numerous intermediate stages that must be involved, except that a reactive C_5 isopentene grouping was implicated. The characterization of the complex reactions concerned in lanosterol biosynthesis together with an understanding of their mechanism has since been almost entirely elucidated by a brilliant series of intensive investigations led by Cornforth and Popják, with equally important contributions by Bloch and also (especially for the early stages) Folkers, Lynen and Rudney.

FORMATION OF MEVALONATE

The chain of reactions leading to the synthesis of cholesterol became clearer when Folkers and associates[10-12] identified (+)-mevalonic acid (III) as a growth factor for acetate-requiring species of *Lactobacilli*. Its structure was later determined as the 6-membered δ-lactone (IV)[13] and confirmed by synthesis. Mevalonate was then tested for precursor activity of cholesterol synthesis and proved to be the most efficient precursor in rat liver

$$\text{OH} \qquad \text{CH}_3$$
$$\text{HOCH}_2.\text{CH}_2.\overset{\displaystyle|}{\text{C}}.\text{CH}_2.\text{CO}_2\text{H}$$

(III)

(IV)

homogenates (43 per cent conversion of the racemic compound).[14] In anaerobic systems in which the hydroxylation step was prevented it was converted into the hydrocarbon squalene.[15] In both cases, the C-1 atom was lost as CO_2.[16] These findings supported the hypothesis that mevalonate acted as the fundamental precursor of the biological isoprenoid unit. Its mode of synthesis has now been established and, for convenience, will be discussed in this section (Scheme 7.1). Other sections will deal with the later stages of sterol biosynthesis.

Preliminary studies[17] showed that rat liver systems prepared from microsomal and soluble enzymes, supplemented with CoASH, ATP, NADPH, glutathione and Mg^{2+}, converted acetate into mevalonate. The initial reaction involved the formation of acetyl-CoA by means of acetokinase (acetyl-CoA synthetase, EC 6.2.1.1) but *in vivo* this would be obtained intramitochondrially from pyruvate or fatty acid oxidation. Presumably acetyl-CoA is then transported to the cytoplasm, at least in animal systems, after conversion into citrate with subsequent cleavage by ATP citrate lyase (EC 4.1.3.8), although this sequence has not been tested directly in connection with isoprenoid biosynthesis. A recent report,[17a] however, does confirm that the synthesis of 3β-hydroxysterols (that is those that are precipitated by the addition of digitonin and including cholesterol) is greatly suppressed in perfused rat livers by hydroxycitrate, an inhibitor of ATP citrate lyase. There is therefore a dependency on the presence of this enzyme for the supply of acetyl-CoA for sterol production in addition to that for fatty acid synthesis (see Chapter 4).

Acetyl-CoA is subjected to the action of acetoacetyl-CoA thiolase (acetyl-CoA:acetyl-CoA C-acetyl transferase, EC 2.3.1.9) to give acetoacetyl-CoA. Next this product reacts with a third molecule of acetyl-CoA under the influence of 3-hydroxy-3-methylglutaryl-CoA (HMG-CoA) synthase [3-hydroxy-3-methylglutaryl-CoA acetoacetyl-CoA lyase (CoA acetylating), EC 4.1.3.5] to generate the C_6 intermediate, HMG-CoA (V),[18] with the elimination of coenzyme A (equation (1) below). An aldol condensation occurs between the methyl group of acetyl-CoA and the carbonyl group of acetoacetyl-CoA as acceptor molecule (in analogy with oxaloacetate in the citrate synthase system). This reaction has been studied by Rudney[18,19] and Lynen[20] with their colleagues and is essentially irreversible. There is, however, a cleavage enzyme present in mitochondria[21] that splits HMG-CoA

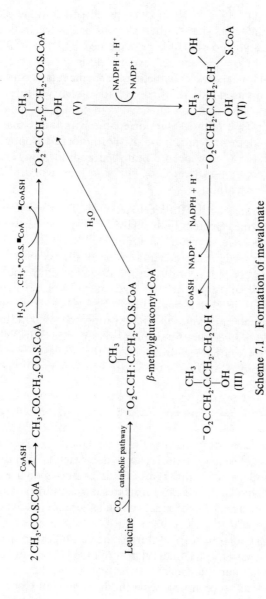

Scheme 7.1   Formation of mevalonate

into acetoacetate and acetyl-CoA (3-hydroxy-3-methylglutaryl-CoA acetoacetate lyase, EC 4.1.3.4) and is concerned with the production of ketone
bodies.[20]

Experiments with [14]C- and ^{3}H-labelled precursors and the yeast enzyme
established that the coenzyme A liberated was generated from acetyl-CoA
while the acetoacetyl thioester linkage remained intact.[22,23] Radioactivity
from [1-^{14}C]acetyl-CoA was located exclusively in the free carboxyl group
of HMG-CoA (V) whereas the ^{3}H-labelled coenzyme A portion was retrieved
as coenzyme A (equation (1)):

$$CH_3.{*}CO.S.{^\bullet}CoA + CH_3.CO.CH_2.CO.S.CoA \xrightarrow{H_2O}$$

$$HO_2{*}C.CH_2.\overset{\overset{\displaystyle CH_3}{|}}{C}(OH).CH_2.CO.S.CoA + {^\bullet}CoASH \quad (1)$$

(V)

This enzyme was also active for acetoacetyl-ACP as substrate, but at a
lower rate, and formed protein-bound HMG.[22] Again, the thioester group
attached to the acetoacetyl group was retained.

We must now consider that the equilibrium of the acetoacetyl-CoA thiolase reaction lies very unfavourably in the direction of synthesis of the C_4
product. This is circumvented in fatty acid and phenol biosynthesis by the
utilization of ATP in the formation of the carboxylated intermediate,
malonyl-CoA. Rudney and colleagues have proposed[23] that the unfavourable
equilibrium might be overcome for isoprenoid biosynthesis by a close orientation of HMG-CoA synthase with respect to the thiolase, coupled with the
irreversible nature of HMG-CoA reductase which is the following enzyme
in the sequence:

$$2\,\text{Acetyl-CoA} \xrightleftharpoons{\text{CoASH}} \text{acetoacetyl-CoA} \xrightarrow{\text{acetyl-CoA}} \text{HMG-CoA} \xrightarrow{\text{NADPH}} \text{mevalonate}$$

However, they have recently shown[24] that a distinct form of acetoacetyl-CoA
thiolase occurs exclusively in the cytoplasmic fraction of yeast cells and
appears to be implicated in this synthesis. It possesses high K_m values for
acetoacetyl-CoA and CoA, the two substrates required by the mitochondrial
enzyme for the reverse cleavage reaction, and therefore favours production
of the β-oxoacyl-CoA.

HMG-CoA may also be formed during the catabolism of leucine *via* the
intermediate formation of 3-methylcrotonyl-CoA and its carboxylated
product, 3-methylglutaconyl-CoA.[25]

The next step in isoprenoid biosynthesis is mediated by HMG-CoA
reductase[26] [mevalonate:NADP oxidoreductase (acylating CoA), EC
1.1.1.34], a microsomal enzyme[27] that shows an absolute requirement for
NADPH (two molecules) for hydrogen transfer to (S)-HMG-CoA. (The

significance of the terms R and S will be covered shortly in the section entitled: 'Stereospecificity in Isoprenoid Biosynthesis'.) The reduction which is essentially irreversible occurs at the thioester group and results in the formation of (R)-$(+)$-mevalonate (III). The only metabolic fate open to this product is that of precursor for isoprenoid compounds. The (S)-form is metabolically inert and the racemic mevalonate may therefore be used as substrate for experimental purposes.[28] Lynen's group[29] recently indicated that mevaldate-CoA hemithioacetal (VI) might act as intermediate in this two-step reaction in yeast. The relative rates of reduction of (VI) and HMG-CoA (V) remained the same during several stages of purification of HMG-CoA reductase. HMG-S-enzyme[29] and free mevaldate (VII),[30] the corresponding aldehyde, were not formed in the course of this reaction.

$$\underset{(VI)}{\overset{\overset{\displaystyle CH_3 \quad OH \quad OH}{\diagdown \diagup \diagup}}{\underset{\diagdown}{HO_2C.CH_2.C.CH_2.CH}}} \qquad \underset{(VII)}{\overset{\overset{\displaystyle CH_3 \quad OH}{\diagdown \diagup}}{HO_2C.CH_2.C.CH_2.CHO}}$$

$$S.CoA$$

Control of Cholesterol Synthesis

The major point at which synthesis of cholesterol is controlled *in vivo* is at HMG-CoA reductase. Early studies on the mechanism of regulation of cholesterol synthesis in liver showed that the level of this enzyme fluctuated greatly under various conditions.[27] The synthesis of HMG-CoA from acetate (in systems deprived of NADPH) and the conversion of mevalonate into cholesterol[31] were not appreciably affected in preparations derived from animals that had been fasted or dosed with high levels of dietary cholesterol. However, the synthesis of mevalonate was markedly inhibited under these conditions. The activity of HMG-CoA reductase was measured directly and was severely reduced to a fraction of normal values. On the basis of these various results, therefore, this enzyme is considered to be rate-limiting in the anabolic sequence leading to cholesterol synthesis.[32] Moreover alternative metabolic pathways are not available for the product of this reaction (mevalonate) and it would seem logical to suppose that this enzyme possesses a regulatory function in the intact cell.

More recently, the reactions leading from acetate to mevalonate in rat liver preparations have been studied individually.[33] Methods were developed that proved more specific in assessing rate-controlling steps and did not rely solely on modifications in the rate of incorporation of radioactive substrates. Supernatant enzymes were capable of converting acetyl-CoA into HMG-CoA but the addition of microsomes was essential for the synthesis of mevalonate. Cholesterol feeding severely depressed the activity of HMG-CoA reductase and to a lesser extent that of HMG-CoA synthase.[34] Previous failure[31] to detect any reduction in the conversion of acetate into

HMG was attributed to the presence of microsomes and hence HMG-CoA reductase (that tended to increase the rate of removal of HMG-CoA).

The mechanism of control is not clearly understood but may be related to the production of bile acids and their effect on HMG-CoA reductase. These are the normal end-products of cholesterol metabolism in the liver and could act as feed-back inhibitors.[31,32] Lynen's group[35] demonstrated stimulation of cholesterol synthesis in rats that were fed with anionic exchange resins to remove bile acids. The results indicated that bile acids together with other factors affected the rate of synthesis and degradation of HMG-CoA reductase. Feedback control by excess cholesterol is totally lost at the level of this enzyme in hepatomas.[36]

FORMATION OF FARNESYL PYROPHOSPHATE

Mevalonate is incorporated into all isoprenoid-derived compounds after loss of the carboxyl group at C-1 and intermediate formation of the C_5-precursor, Δ^3-isopentenyl pyrophosphate (isopentenyl-PP, (VIII)). Coenzyme

$$CH_3.C(:CH_2).CH_2.CH_2.O.\overset{\overset{O}{\|}}{\underset{\underset{O^-}{|}}{P}}.O.\overset{\overset{O}{\|}}{\underset{\underset{O^-}{|}}{P}}.OH$$

(VIII)

A derivatives are involved up to this stage in the synthesis of isoprenoids but are not required for the reactions subsequently engaged by mevalonate. Early experiments with [2-^{14}C]mevalonate readily established that the squalene formed did not possess radioactivity in the branched methyl groups. Thus the identity of C-2 remained distinct after loss of the carboxyl group and it did not become equivalent to C-3′ in these positions. The reaction sequence by which mevalonate is converted into farnesyl-PP has now been established and all the enzymes concerned have been isolated from high-speed supernatant fractions from liver and characterized (Scheme 7.2).

Derivation of Isopentenyl-PP

The initial two reactions are successive phosphorylations. The first is catalysed by mevalonate kinase (ATP:mevalonate-5-phosphotransferase, EC 2.7.1.36)[13] and involves the esterification of the primary hydroxyl group to give 5-phosphomevalonate. This enzyme is specific for the dextrorotatory (3R)-enantiomer [but is inactive for the lactone (IV)] and requires divalent ions. It is strongly inhibited by p-chloromercuribenzoate which binds to essential thiol groups but this effect may be overcome by the addition of glutathione. The product is then subjected to the action of phosphomevalonate kinase (ATP:5-phosphomevalonate phosphotransferase, EC 2.7.4.2) to form 5-pyrophosphomevalonate (mevalonate-5-PP), with hydrolysis of the terminal phosphoryl group of a second molecule of ATP.

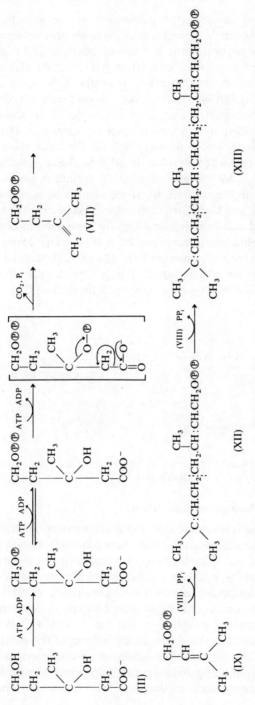

Scheme 7.2 Formation of farnesyl-PP (XIII) from mevalonate. The dotted lines divide the molecules into isoprenoid-derived units; PP and PP_i denote $HP_2O_6^{2-}$ and $HP_2O_7^{3-}$ respectively

The next reaction involves the conversion of mevalonate-5-PP into Δ^3-isopentenyl-PP through the agency of pyrophosphomevalonate decarboxylase [ATP:5-pyrophosphomevalonate carboxylase (dehydrating), EC 4.1.1.33]. In the course of the decarboxylation ADP, P_i and CO_2 are liberated (and ATP consumed) in stoichiometric amounts. It is possible that this decarboxylation–dehydration reaction proceeds *via* the intermediate formation of the tertiary phosphate (as depicted in Scheme 7.2) since the hydroxyl oxygen atom is located in the orthophosphate formed.[37] However, the postulated intermediate, 3-phosphomevalonate-5-PP, has not been detected in liver or yeast systems when mevalonate-5-PP has been presented with the purified enzymes.[38,39] Moreover, deuterium or tritium is not incorporated into the product from the medium when the incubation is performed with D_2O or T_2O. Popják and Cornforth,[38] therefore, favour a concerted mechanism in accord with this information in which the terminal phosphate of ATP is displaced on acceptance of electrons from the tertiary hydroxyl group, simultaneously with the elimination of CO_2, to give isopentenyl-PP (Scheme 7.3). The two hydrogen atoms attached to C-2 of mevalonate-5-PP are retained in the terminal methylene group of this product.

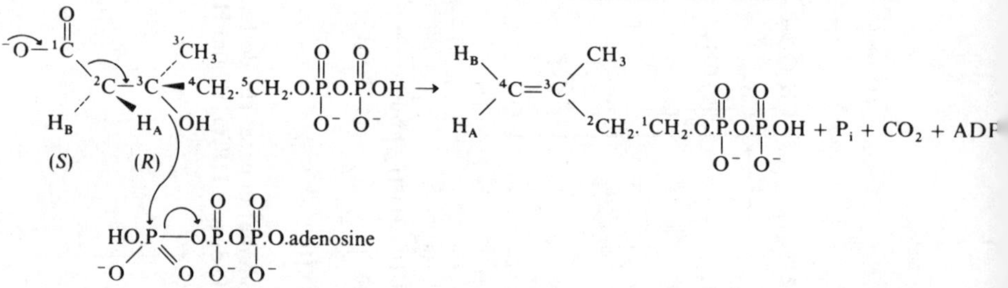

Scheme 7.3　Stereochemical elimination of the carboxyl and tertiary hydroxyl groups of mevalonate-5-PP[39]

Stereospecificity in Isoprenoid Biosynthesis

Although squalene is a symmetrical molecule, its structure hides the fact that its formation includes many asymmetrical processes. Popják and Cornforth[38] have pointed out that there are 14 possible sites in the overall conversion of mevalonate to squalene (including modifications at a methylene group) at which stereospecific features may occur, despite the lack of optical asymmetry in the products. This behaviour therefore is highly reminiscent of the different reactivity exhibited by the two $-CH_2.CO_2H$ groupings in citrate which depends on the asymmetrical binding of this substrate to the surface of the aconitase enzyme. The active sites of the enzymes concerned are capable of differentiating between two apparently identical functions by virtue of their asymmetric nature, and exhibit precise stereochemical control. These reactions have been extensively studied by means of elegantly prepared experiments in which labelled substrates including NADPH were synthesized stereospecifically.[40–42] Their fates were examined after incubation with purified enzymes and the products analysed by sophisticated techniques including mass spectrometry and gas–liquid radiochromatography. Knowledge of the stereospecific direction of any reaction is important because it imparts detailed information concerning the manner in which that reaction proceeds. Mevalonate has been the principal substrate used in these investigations and has been synthesized by as many as 19 distinct reaction sequences[43] dependent on the ^{14}C-, ^{2}H- or ^{3}H-labelling requirements, including asymmetrical labelling at the three methylene groups with ^{2}H or ^{3}H.[38–42]

In order to appreciate the significance of the results gained from these experiments, we must first consider the meaning of the R and S nomenclature [Latin for *rectus* (right), *sinister* (left)] for designation of chirality. This term refers to stereoisomers which are not superimposable with their mirror images and reflects the asymmetry of the molecule. Notation of R and S is determined unambiguously by the sequence rules of Cahn, Ingold and Prelog[44] and is based on the order of substituents in decreasing size of the atomic number (or if this is identical, the atomic weight) of the atoms attached to the carbon atom concerned, when viewed from the side remote from the smallest substituent (often H). This convention provides a standard point of reference for each pair of isomers. The central carbon atom, as examined in a model, would then be seen plus three substituents lying in one plane in a given sequence. Thus ^{18}O takes precedence over ^{16}O and D over H; the priority for a number of atoms is therefore in the order

$$S > P > O > N > C > T > D > H.$$

If this assessment does not prove decisive, a similar argument is applied in terms of the priority accorded to subsequent atoms within the groups, e.g.

$$CH_2OH \begin{bmatrix} & OH \\ & / \\ C{-}H \\ & \backslash \\ & H \end{bmatrix} \text{ precedes } CH_3.$$

Now consider the molecule Ca_1a_2bd where 'a' represents two identical groups or hydrogen atoms attached to two valencies of the prochiral centre (carbon atom) in question. Each group 'a' is placed in a different enantiomeric environment. The convention of Cahn et al.[44] designates the stereoisomers of this molecule as R or S when replacement of an 'a' group (that is H in the situations usually dealt with in the context of isoprenoid biosynthesis) with the heavy isotope, deuterium (^{2}H) or tritium (^{3}H), confers R or S chirality. This is illustrated in the examples chosen below for the mono-deuterated $[4\text{-}^2H_1]$mevalonate in which (a) represents the R stereoisomer with respect to the configuration around C-4 while (b) refers to the S stereoisomer (as follows):

∴ 4 R-isomer (right-handed; clockwise movement)

∴ 4 S-isomer (left-handed; anti-clockwise movement)

The stereospecific nature of the reactions engaged by suitably prepared species of mevalonate and later intermediates may therefore be demonstrated if these substrates lose or the product gains the isotopic label. The application of the techniques devised by Cornforth and Popják with their collaborators has provided answers to all the stereospecific ambiguities except one. The problem concerning the addition of a proton to C-4 of isopentenyl-PP in the formation of dimethylallyl-PP has proved the most difficult and has not yet been clarified. The ability to discriminate between the three isotopes of hydrogen attached to the terminal carbon atom would be required for this purpose.

To return now to a discussion of the first of the individual reactions concerned, it has been established by means of stereospecifically labelled substrates that the decarboxylation of mevalonate-5-PP involves a *trans*-elimination of the carboxyl and tertiary hydroxyl groups. The reaction results in the *pro-R* hydrogen atom originally attached to C-2 of mevalonate (H_A) bearing a *cis*-orientation to the pyrophosphoryl substituent at C-3 of isopentenyl-PP (Scheme 7.3).[39] It should be clarified at this point that the numbering system in mevalonate differs from that in isopentenyl-PP; the carbon atoms in both compounds are numbered in Scheme 7.3. This conversion is analogous to the concerted decarboxylation–debromination of β-bromocarboxylates that leads to olefinic products by means of a *trans*-elimination of CO_2 and bromide ion.[45] Information concerning its stereospecificity was obtained after studying the products of bromination and dehydrobromination of the isopentenol (derived from $[2R\text{-}2\text{-}^2H_1]$- and $[2S\text{-}2\text{-}^2H_1]$mevalonate) that was liberated on treatment of isopentenyl-PP with alkaline phosphatase.[38] Isopentenyl-PP accumulated in the incubation medium and was therefore available for study after addition of iodoacetamide to the enzyme preparation, when isomerization was prevented. Thus the deuterium atom present in the 2R-configuration in [2-²H₁] mevalonate, defined as H_A in Scheme 7.3, was retained in the product in the *cis* position relative to the larger substituent.

Stereospecifically labelled [2-²H₁]mevalonate was prepared[39,42] for this and other experiments by a chemical interchange of the terminal functional groups of [4R-4-²H₁]- and [4S-4-²H₁]mevalonate. Mixed enantiomers of 3R, 4R-, 3S, 4S-, 3R, 4S- and 3S, 4R-mevalonate were used as starting material. The carboxyl groups were initially protected by conversion into their methyl esters when the primary alcohol groups were oxidized through to the acids. Finally, the methyl esters were reduced to the alcohol stage. Thus C-5 and C-4 in the original mevalonate become C-1 and C-2 in the modified product while stereospecifically placed hydrogen atoms generate labels of opposite chirality. There is also therefore a resultant inversion of configuration around C-3 (Scheme 7.4) but this does not interfere with the experimental arrangements since only the 3R-enantiomer reacts with mevalonate kinase.

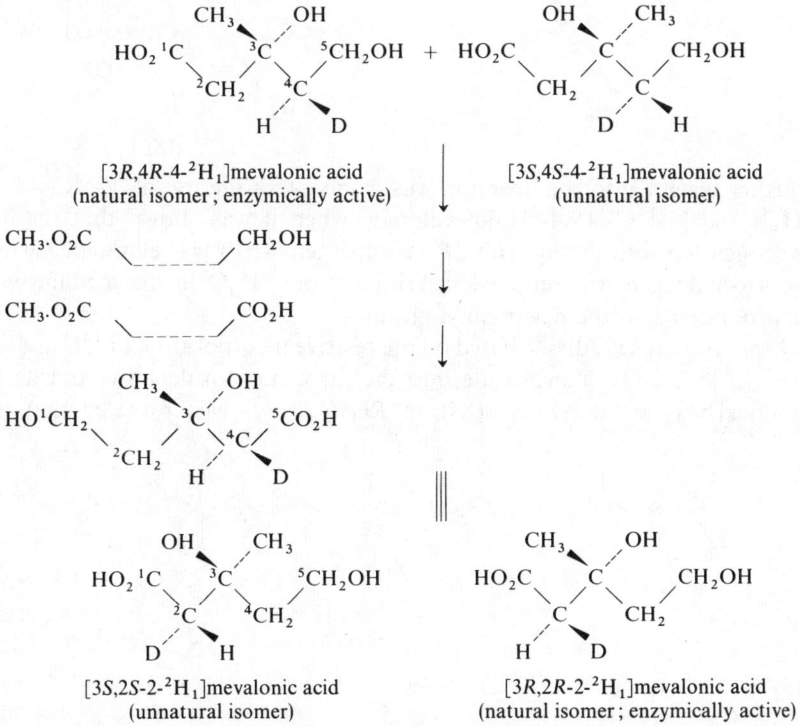

[3R,4R-4-²H₁]mevalonic acid
(natural isomer; enzymically active)

[3S,4S-4-²H₁]mevalonic acid
(unnatural isomer)

[3S,2S-2-²H₁]mevalonic acid
(unnatural isomer)

[3R,2R-2-²H₁]mevalonic acid
(natural isomer; enzymically active)

Scheme 7.4   Chemical conversion of racemic [4-²H₁]mevalonic acid into racemic [2-²H₁]mevalonic acid

Chain Elongation

Isopentenyl-PP next condenses with its reactive isomer, 3,3-dimethylallyl-PP (IX), in a process of chain elongation.[38] The isomerase (isopentenyl-PP isomerase, EC 5.3.3.2) reaction is reversible but is directed towards the formation of the allyl product (IX) (90 per cent at equilibrium) in yeast[46] and liver[47] enzyme. Lynen[48] has described the chemical consequences of this isomerization as the conversion of isopentenyl-PP, containing an unreactive phosphoryl group and nucleophilic double bond, into a highly reactive electrophilic substrate (IX). The reaction proceeds *via* an enzyme-bound intermediate (possibly a thioether) and is inhibited by thiol group reagents.[46] It is initiated by proton attack on the methylene carbon atom,[47] followed by stabilization of the transitory carbonium ion formed by loss of a proton from C-2 (equation (2)):

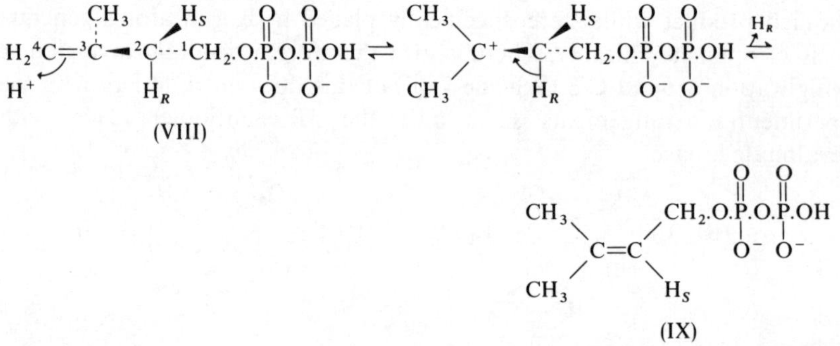

$$H_2{}^4C={}^3C-{}^2C\cdots{}^1CH_2.O.\overset{O}{\overset{||}{P}}.O.\overset{O}{\overset{||}{P}}.OH \rightleftharpoons$$

(VIII)

$$\rightleftharpoons$$

$$\underset{CH_3}{\overset{CH_3}{\diagdown}}C=C\underset{H_S}{\overset{CH_2.O.\overset{O}{\overset{||}{P}}.O.\overset{O}{\overset{||}{P}}.OH}{\diagup}} \quad (2)$$

(IX)

Further insight into this reaction was gained with the use of [2-^{14}C, 4R-4-^{2}H$_1$]- and [2-^{14}C, 4S-4-^{2}H$_1$]mevalonate when it was shown that the 4S-hydrogen (corresponding with 2R in isopentenyl-PP) was eliminated while the 4R-hydrogen was retained.[42] Tritium from ^{3}H$_2$O in the medium was incorporated into the new methyl group.

More recently, studies[49] based on the relative incorporation of ^{14}C and ^{3}H from [2-^{14}C, 2-^{3}H$_2$]mevalonate into the carotenes, torulene (X) and its 1'-monocarboxylate derivative (XI), in *Rhodotorula* yeast revealed that the

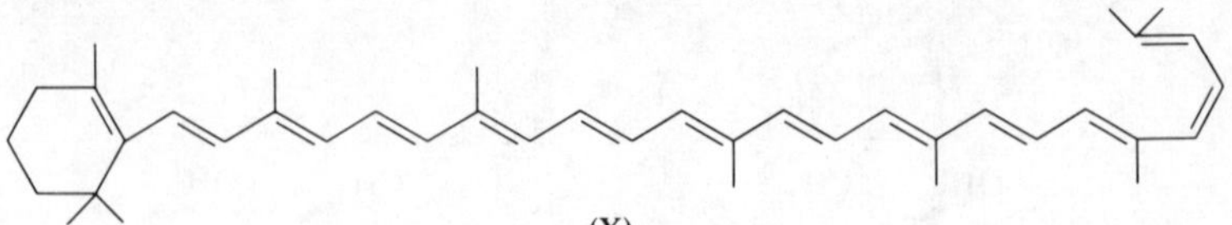

(X)

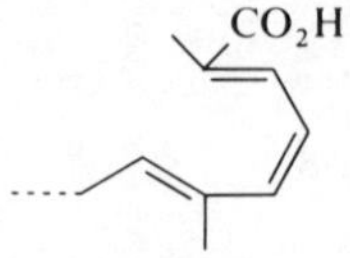

(XI)

methyl groups within the *gem*-dimethyl unit (derived initially from C-2 and C-3′ of mevalonate and formed by isomerization of isopentenyl-PP) retained their individuality. Accordingly, a modified mechanism for the isomerase reaction was formulated in which a concerted attack by a proton and thiol group of the enzyme occurred which prevented randomization (equation (3)):

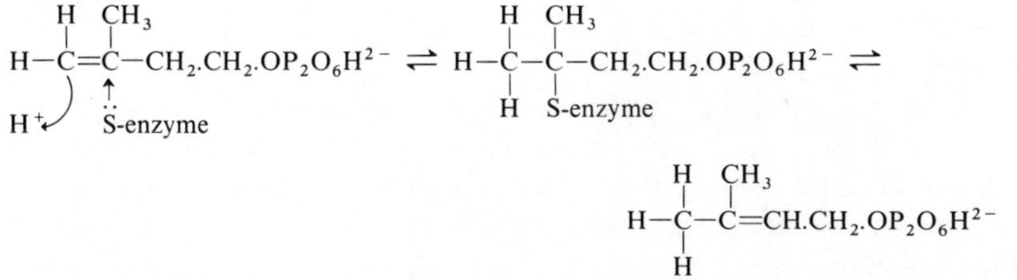

$$
\underset{\substack{H^+ \\ }}{\overset{\substack{H \quad CH_3}}{H-C=C-CH_2.CH_2.OP_2O_6H^{2-}}} \overset{\substack{}}{\rightleftharpoons} \underset{\substack{H \quad S\text{-enzyme}}}{\overset{\substack{H \quad CH_3}}{H-C-C-CH_2.CH_2.OP_2O_6H^{2-}}} \rightleftharpoons
$$

$$
\underset{H}{\overset{\substack{H \quad CH_3}}{H-C-C=CH.CH_2.OP_2O_6H^{2-}}} \quad (3)
$$

Isopentenyl-PP and the isomeric dimethylallyl-PP both act as substrates for the condensation reaction, catalysed by a soluble prenyl transferase (geranyltransferase, EC 2.5.1.1), in which a proton and pyrophosphate group are lost from (VIII) and (IX) respectively.[48] The product is the C_{10} geranyl-PP (XII) which possesses the reactive allyl pyrophosphate grouping and is capable of further elongation in an analogous manner by association with a second molecule of isopentenyl-PP and probably the same transferase enzyme. The C_{15} unit, farnesyl-PP (XIII), is thereby formed (Scheme 7.5).

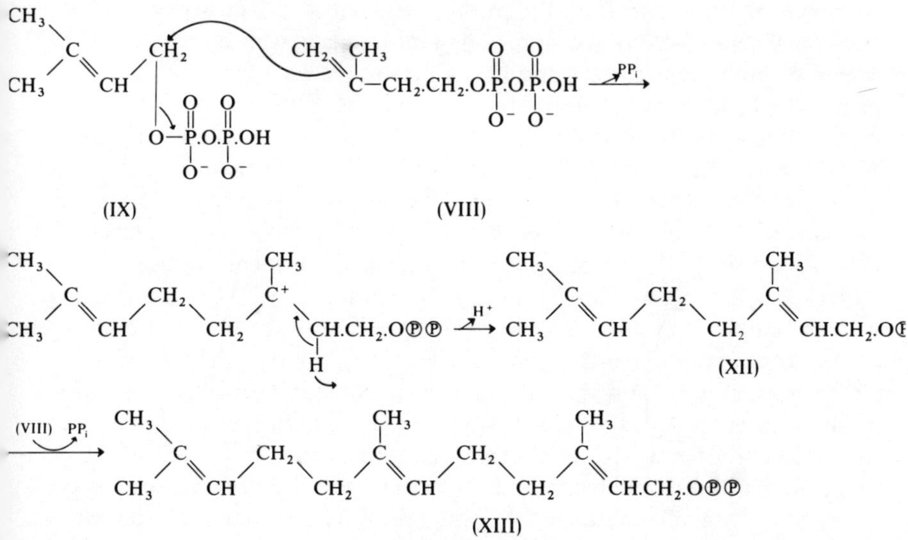

Scheme 7.5   Formation of farnesyl-PP from isopentenyl-PP and dimethylallyl-PP

The enzyme concerned was first recognized in yeast preparations[48] but has since been isolated from pig liver.[50,51] It catalyses both stages but is unable to utilize farnesyl-PP as a further substrate. However, other polyprenyl transferase enzymes do exist in Nature with more general specificities, e.g. those responsible for the formation of geranylgeranyl-PP (C_{20}), a precursor of carotenoids and vitamin A, and for the synthesis of the polyisoprenoid rubber and gutta (Chapter 8). The isolation of ubiquinone, vitamin K_2, etc. and long-chain alcohols clearly indicates the enzymic capacity for their synthesis.

Holloway and Popják[51] purified the dimethylallyl and geranyl trans-ferase(s) from pig liver homogenates but even after extensive purification of enzyme the ratio between their activities remained constant. The products were characterized after treatment with alkaline phosphatase and subjection of the alcohol to analysis by gas–liquid radiochromatography. In both cases the major fraction consisted of *trans-trans*-farnesol, a feature that indicated a stereospecific course in the transferase reactions. Kinetic data and analysis of reaction mechanisms related to the order of binding of substrates and subsequent release of products were obtained from product inhibition studies using geranyl-PP and isopentenyl-PP as substrates.

These transferase reactions were also studied with asymmetrically-prepared ^{2}H- and ^{3}H-labelled substrates. Examination[38] by mass spectrometry of the farnesol obtained from [$4R$-4-^{2}H$_1$]mevalonate showed the presence of the trideuterated species as it gave a molecular ion in agreement with $M + 3$ where M corresponds to the molecular weight. Further results confirmed that protons from the *pro-R* hydrogen at C-2 of isopentenyl-PP and C-6 of geranyl-PP (*pro-S* from C-4 of mevalonate) were eliminated in the course of both transferase reactions as occurs in the isomerization of iso-pentenyl-PP. Moreover, information obtained with the use of [$5R$-5-^{2}H$_1$]-mevalonate demonstrated that inversion of configuration occurred around the allylic carbon atom of dimethylallyl-PP,[41,42] on formation of the new C—C bond (Scheme 7.6). With regard to the steric position of hydrogen attached to C-4 of geranyl-PP (or the corresponding C-4 and C-8 of farnesyl-PP), using [$2R$-2-^{2}H$_1$]mevalonate as substrate, the evidence indicated that dimethylallyl or geranyl allylic groups formed this C—C bond by reacting with the vinylic carbon (C-4) of isopentenyl-PP from the side on which the groups appeared in the order: —CH$_2$.CH$_2$.OPP, —CH$_3$, —H and —H in a clockwise manner.[39] Further, the data was compatible with the reaction mechanism entailing two-steps, firstly the *trans* addition of the allylic unit from (IX) or (XII) and also an electron donating group, X$^-$ (attached to C-3) to the double bond (from below and above the plane of the bond respectively), followed by *trans* elimination of X$^-$ plus a proton (H$_R$) from C-2 of the inter-mediate (XIV) (Scheme 7.6). The inversion at C-1 is characteristic of a bi-molecular nucleophilic substitution (S_N2 displacement). The identity of X

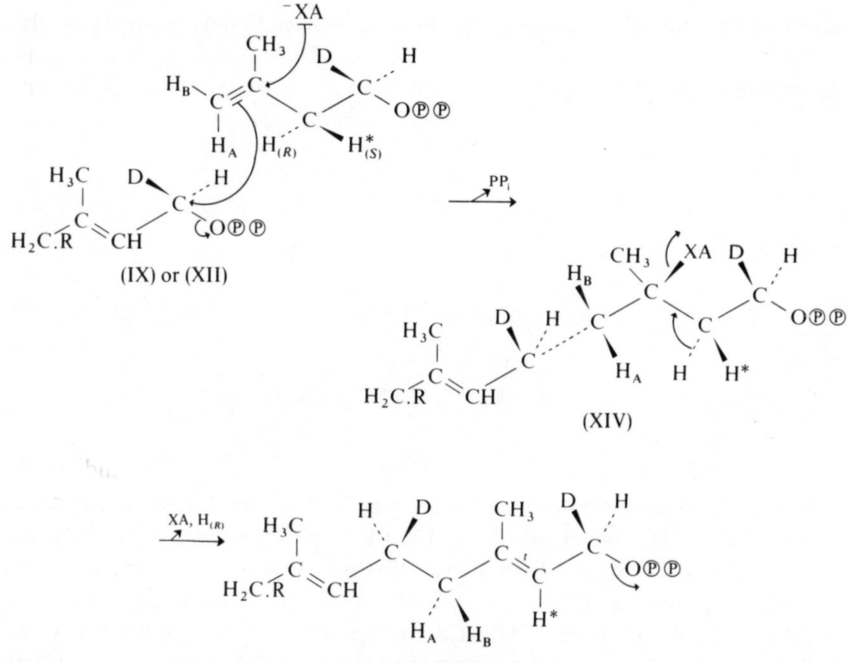

Scheme 7.6 Stereospecificities involved in the transferase reactions. R = H— or (CH$_3$)$_2$.C:CH.CH$_2$— (from Popják and Cornforth, 1966)[38]

has not been settled[52] but it may belong to a group on the enzyme (possibly a sulphydryl anion), or the pyrophosphate ion of isopentenyl-PP may serve this function. Cornforth et al.[39] have further speculated that X may also be implicated in the isomerase reaction (equation (2)) since the addition of a proton to the vinylic carbon of isopentenyl-PP is analogous to that of the allylic group.

SYNTHESIS OF SQUALENE

The biosynthesis of isoprenoid compounds with the exception of mono-terpenes (C$_{10}$) share the same reaction sequences as far as farnesyl-PP. This product is then utilized in many divergent pathways for synthetic purposes; some of these will be dealt with in Chapter 8. Lynen originally showed that yeast extracts could synthesize squalene (XV), a C$_{30}$ triterpenoid hydrocarbon, anaerobically in the presence of NADPH; if absent, farnesyl-PP accumulated. The reaction leading to squalene from two molecules of farnesyl-PP (its immediate precursor) follows a completely different course from the previous condensations and involves the formation of a C—C bond between C-1 of each reacting substrate. In the process, two pyrophosphate groups are released with loss of a hydrogen atom from one molecule of farnesyl-PP and replacement by H derived from NADPH[53] (equation (4)):

(XIII)

$2^-HO_6P_2O$

(XIII)

NADPH   NADP$^+$ 2PP$_i$

(4)

(XV)

Information concerning this reaction was first revealed with the use of the deuterated [5-$^2H_2$]mevalonate and doubly labelled [1-$^2H_2$, 2-^{14}C]farnesyl-PP.[53,54] Squalene formed from these substrates was subjected to ozonolysis and the resulting succinate was examined by mass spectrometry after conversion into the anhydride. This fragment contained the grouping —CD$_2$.CHD— which was present at the centre of the squalene molecule, clearly indicating that only one of the original deuterium atoms had been lost in the condensation reaction. Experiments that utilized double labelling techniques proved extremely valuable in according confirmatory data but before this approach is described a short explanation for their rationale will be presented.

The radioactivity residing in a sample of doubly-labelled substrate is accurately determined after counting in a two channel scintillation spectrometer and correction for overlap in the counts given by the two isotopes (e.g. ^{3}H and ^{14}C) and quenching. The ratio of the counts observed is normalized to give a fixed relative value, say 1:1 for the ^{3}H/^{14}C atomic ratio. If four tritium atoms are involved in the reaction to be studied as in the condensation between two ditritiated substrates, a working ratio of 4:4 may prove more convenient for the expression of results. The ^{14}C-label in the substrate is used to give an internal reference. The labelled product must be rigidly purified before assay of the radioactivity generated by each isotope. Even trace contamination by an additional product with similar properties but containing a different ratio could distort the results and render the analysis meaningless. If one tritium atom (from an original total of four) has been lost during the reaction process, this would be reflected in the reduced ratio of approximately 0·75:1 (3:4) for ^{3}H/^{14}C that is obtained. Ideally, the substrate presented for these investigations should be doubly labelled internally within the molecule that is [^{14}C, ^{3}H]substrate rather than a mixture of the individual

[^{14}C]- and [^{3}H]substrates, to avoid problems arising from possible discrimination against the tritiated substance because of the large relative difference in atomic weight between ^{3}H and H.

Returning now to squalene formation, incubation of squalene synthetase with [1-^{3}H$_2$, 2-^{14}C]farnesyl-PP (with a normalized ratio for ^{3}H/^{14}C as 1:1) resulted in the production of squalene with an atomic ratio of 0.76:1, clearly showing that a central tritium atom had been lost.[55] Since this value indicated that exactly 1 in 4 of the tritium atoms had been lost, there was no discrimination against the heavy isotope from the terminal —C^{3}H$_2$.O. P$_2$O$_6$H^{2-} grouping compared with the normal —CH$_2$.O.P$_2$O$_6$H^{2-}. Moreover, farnesyl-PP that was biosynthetically prepared from [2-^{14}C, 5-^{2}H$_2$]mevalonate contained the hexadeuterated 1,5,9-^{2}H$_6$-labelled isomer, as determined by its molecular ion of $M + 6$.[55] Experiments involving the use of NADP^{3}H and ^{3}H$_2$O demonstrated conclusively that NADPH was the source of the replaced hydrogen atom in squalene.[53]

Many possible mechanisms have been presented for the squalene synthetase reaction but real progress into its intimate details was first made with studies using stereospecifically labelled substrates. It had previously been established[56] that NAD$^+$- and NADP$^+$-linked oxidoreductase enzymes could be classified into two groups, A and B, that differed from each other in the particular hydrogen they transferred from C-4 of the reduced nicotinamide ring to the substrate. Popják and Cornforth[38,40] prepared both ^{3}H-labelled specimens of NADPH by reduction of [4-^{3}H]NADP$^+$ with glucose 6-phosphate dehydrogenase (EC 1.1.1.49, class A enzyme) and isocitrate dehydrogenase (EC 1.1.1.42, class B enzyme), together with their appropriate substrates. They identified squalene synthetase as belonging to class B as it only transferred tritium from NADP^{3}H which had been derived from isocitrate. Moreover, these workers determined the absolute configuration of these positions relative to C-4 of the ring and classified the 'A' and 'B' sides as the 4*pro-R* and 4*pro-S* hydrogen respectively, by reference to the known stereochemistry of *S*-malate [produced by fumarate hydratase (EC 4.2.1.2)].[57] Thus the 4*pro-S* hydrogen is transferred as a hydride ion to one of the farnesyl residues in the synthesis of squalene.

Experiments with the hexadeuterated farnesyl-PP (formed from [5-^{2}H$_2$]-mevalonate) and NADPH showed that the deuterium atom was eliminated from the 1*S* position of a farnesyl-PP molecule.[58] The squalene formed was subjected to ozonolysis and the resulting trideuterated succinic acid obtained from the central carbon atoms, was identified after analysis of its optical-rotatory-dispersion curve and comparison with the authentic mono-deuterated [2*R*-2-^{2}H$_1$]succinic acid which was used as reference material. Its identity proved to be [2*S*-2-^{2}H$_1$, 3-^{2}H$_2$]succinic acid (XVI), giving the structure (XVII) to squalene (Scheme 7.7). The configuration at C-1 of the farnesyl group that does not undergo exchange is inverted on formation of

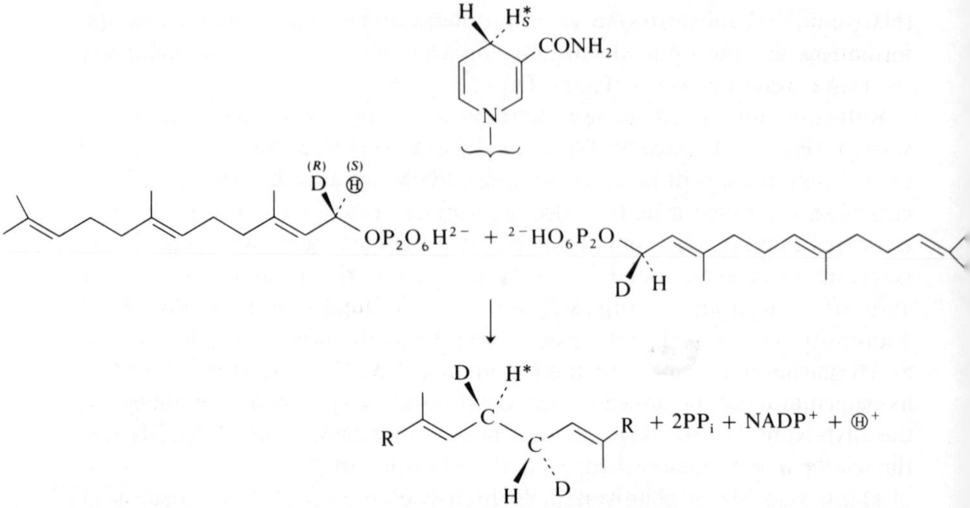

Scheme 7.7   Stereochemistry of the formation of squalene from farnesyl-PP and NADPH (Popják, 1970)[58]

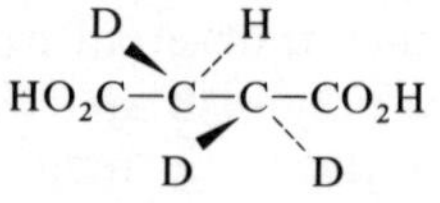

$$HO_2C-C-C-CO_2H$$

(XVI)

the new C—C bond (Scheme 7.7). These results were independently confirmed in experiments using stereospecifically prepared substrates.[38] When [5R-5-$^2H_1$]mevalonate was provided, all the deuterium atoms were retained to give [$^2H_6$]squalene. In addition, squalene and farnesyl-PP derived from [4-^{14}C, 5R-5-$^3H_1$]mevalonate both gave the same $^3H/^{14}$C atomic ratio, indicating that the hydrogen atom removed during the condensation step arose from the 1pro-S position of farnesyl-PP. The 5S-labelled mevalonate has not been prepared to test directly since the 5R-enantiomer is formed enzymically by reduction of mevaldate with [4-$^2H_1$]- or [4-$^3H_1$]NADH.

   Various suggestions have been made at different times with respect to the mechanism of action of the synthetase in liver and yeast systems. The principal mode of attack has been to study the nature of products that accumulate in an incubation mixture containing farnesyl-PP and enzyme but lacking NADPH, thereby blocking the reductive step. Rilling[59] showed with a particulate preparation from yeast that intermediates did indeed build up. A precursor was isolated and then converted into squalene in a fresh system to which NADPH had been added, and also on incubation with rat liver microsomes plus NADPH. It retained half the initial ^{32}P-label, relative to ^{14}C, indicating that one pyrophosphate group had been lost in its formation. In addition, one of the four tritium atoms originating from $[1-^3H_2]$farnesyl-PP was eliminated at this stage.

Rilling,[60] Popják[61] and their colleagues examined the properties of this stable precursor, named presqualene-PP, following reduction or enzymic hydrolysis to the corresponding C_{30} monohydric alcohol. This derivative is simpler to work with than the pyrophosphate ester. Investigations by n.m.r. and mass spectrometry and catalytic hydrogenation indicated a structure containing 5 double bonds with molecular formula $C_{30}H_{49}OH$ and possessing a tetrasubstituted cyclopropane ring (XVIII). The elemental composition $C_{30}H_{46}D_3OH$ was given by material formed in the presence of $[1-^2H_2]$farnesyl-PP and unambiguously demonstrated the loss of one hydrogen atom at the level of synthesis of presqualene-PP. It is thought that the allylic double bond of farnesyl-PP is initially attacked at C-2 by a nucleophilic site within the enzyme; this carbon atom could then react with C-1 of a second molecule of farnesyl-PP with loss of pyrophosphate, by means of an S_N2 displacement. Next, a proton is released to generate the cyclopropane ring. There is a resultant inversion of configuration around C-1 of the second substrate molecule. Finally, squalene is formed after reduction with the H_S hydride ion from NADPH. The latter process may proceed on conversion of presqualene-PP into a suitable form by ring expansion and migration of the pyrophosphate group[61] to structure (XIX) or by concerted loss of pyrophosphate with rearrangement.[60]

(XVIII)   (XIX)

The formation of nerolidyl-PP (XX), an isomer of farnesyl-PP that contains a tertiary pyrophosphate group has previously been implicated[53] as

(XX)

one of the condensing residues but does not appear to be involved.[62] It has also been suggested that enzyme-bound intermediates might be concerned in squalene biosynthesis[42] and there has been a report on work with a soluble microsomal system from pig liver that appeared to confirm this. However the C_{15}- and C_{30}-enzyme intermediates that were isolated were later shown to be artifacts.[63]

CYCLIZATION OF SQUALENE

The biosynthesis of cholesterol (I) from squalene involves the intermediacy of lanosterol (4,4,14α-trimethylcholesta-8,24-dien-3β-ol, (XXI)), a product

(XXI)

with a fused tetracyclic ring system plus a C_8 side-chain, by means of a cyclization and concerted rearrangement.[8,9] (β-Configurations are designated those that lie above the plane of the ring and are expressed as complete lines in the appropriate formulae.) In this process, a β-hydroxyl group is inserted at C-3 while two intramolecular 1,2-methyl shifts and two 1,2-hydrogen shifts occur.[64] A proton is eliminated. Indeed, Ruzicka[65] proposed a 'biogenetic isoprene rule' rationalizing the cyclization process and proposing that squalene and related aliphatic compounds were the precursors of cyclic terpenoids by modifications of this sequence. In the case of the 'hydroxylase' responsible for lanosterol synthesis, the squalene chain assumes a chair-boat-chair-boat-unfolded conformation that is accepted by that particular enzyme. Moreover, nearly all these polycyclic products possess a 3β-hydroxyl group and exhibit a pattern of folding that is indicative of cyclization induced by a cationic attack at C-3.[66] The oxygen atom is derived from atmospheric O_2 as evidenced by work with liver preparations using $^{18}O_2$ and $H_2^{18}O$ as tracer substrates.[67] Deuterium from ^2H_2O is not incorporated into lanosterol. These results therefore supported Ruzicka's proposition that a species such as OH^+ might prove to be the initiating agent. ^{13}C-labelling studies by Bloch[68] and Cornforth[69] and their colleagues demonstrated conclusively that the methyl-group migrations from squalene were accomplished by two 1,2-shifts rather than a single 1,3-shift.

Further information concerning the details of this reaction awaited the independent proposal by the Laboratories of van Tamelen and Clayton,[70] and Corey[71,72] that the 'squalene oxidocyclase' system might actually be a two-step process catalysed by distinct enzymes. They investigated the possibility that all-*trans*-squalene was first converted into its 2,3-epoxide (XXII) by a monooxygenase reaction involving O_2 and NADPH, followed by cyclization of this product after a proton attack on the epoxide ring, to yield an intermediate carbonium ion (or enzyme-bound product) with an electron deficiency at C-20. Lanosterol would then be formed after stabilization by methyl group and hydrogen transfers, and release of a proton from C-9 (Scheme 7.8). This consideration was based on experiments with acid-catalysed (non-enzymic), model systems they had devised with analogous oxides.[73]

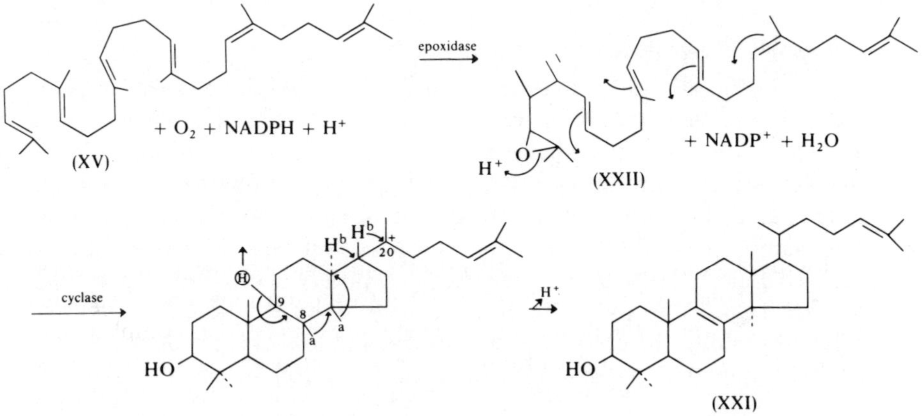

Scheme 7.8 Epoxidation, cyclization and rearrangements involved in the biosynthesis of lanosterol. (a) Migration of methyl groups; (b) migration of hydride ions

The validity of an epoxidation mechanism was substantiated by these workers using rat liver preparations.[70–72] They reported the formation of radioactive squalene-2,3-oxide after incubation with ^{3}H-labelled squalene, NADPH and an exogenous supply of the oxide as carrier. Its addition resulted in a corresponding dilution of incorporation into total sterols. More directly, ^{14}C-labelled squalene-2,3-oxide was converted anaerobically into lanosterol. The major intracellular site of cyclase activity proved to be the microsomes. Tritium and ^{18}O from doubly labelled squalene-2,3-oxide were both effectively incorporated into lanosterol by this fraction, with quantitative retention of ^{18}O, by a process that was independent of NADPH.

A solubilized preparation of squalene-2,3-oxide cyclase was achieved by treatment of hog liver microsomes with deoxycholate in the presence of buffers of high ionic strength.[74] The purified enzyme gave lanosterol as the only product after isomerization. The yeast enzyme behaved similarly but was located in the non-membranous region of the cells.[75]

A selective difference in the properties of the epoxidase and cyclase enzymes has enabled them to be separated from each other. Heat treatment of rat liver microsomes removed cyclase activity but left the capacity for epoxidation unaffected.[76] Accordingly these microsomes accumulate the 2,3-oxide on incubation with squalene since conversion into lanosterol is prevented. The dependency for NADPH and O_2 occurs at the epoxidase level and there is also an absolute requirement for a supplementary supernatant fraction. Thus the involvement of two enzymes in the conversion of squalene into lanosterol has been fully authenticated. Corey's group[71] have speculated that the cyclization process may be initiated by a protonated species within the enzyme.

Conversion of Squalene into Cycloartenol (XXIII)

A major question in connection with the biosynthesis of sterols in plant tissues lies in the nature of the product formed on cyclization of squalene-2, 3-oxide. A great deal of evidence has been collated by Goad[77] that indicates that lanosterol is not implicated in the biogenesis of phytosterols. On the other hand, the closely related triterpene, cycloartenol (XXIII), occupies this central position in the plant kingdom. Squalene-2,3-oxide may be cyclized anaerobically to this precursor in a variety of cell-free preparations derived from plants and algae. In the course of this reaction, there are two 1,2-methyl shifts, three 1,2-hydride transfers and loss of the proton from the methyl group at C-10 which generates the cyclopropane ring (Scheme 7.9).

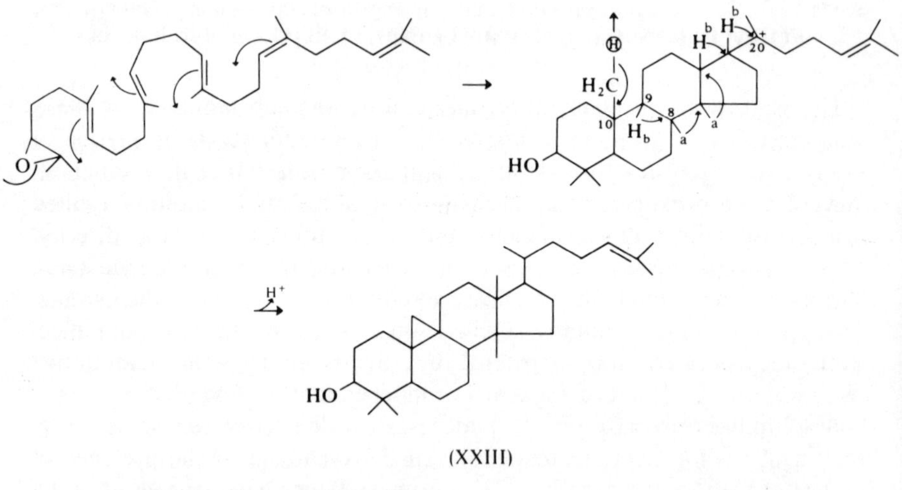

(XXIII)

Scheme 7.9   Formation of cycloartenol from squalene-2,3-oxide. (a) Migration of methyl groups; (b) migration of hydride ions

Biosynthesis of Tetrahymanol (XXIV)

The protozoan genus *Tetrahymena* has been widely used in many bio-chemical studies and is particularly notable for its ability to synthesize a polycyclic terpenoid ring system without the intervention of oxygen. Cell-free preparations of *Tetrahymena pyriformis* convert squalene anaerobically into the pentacyclic triterpene, tetrahymanol (XXIV)[78] which occurs in membrane lipids as a complex bound to phospholipids. The hydroxyl group at C-21 and the hydrogen at C-3 in this metabolite are both derived from water as evidenced by experiments in which $H_2{}^{18}O$ and D_2O were used as tracers. Squalene-2,3-oxide has also been excluded as a precursor confirming the lack of requirement for molecular O_2. The mechanism proposed for this cyclization (and possibly that responsible for the formation of other 3-deoxy triterpenes) is one mediated by proton attack at C-3 with concerted gain of a hydroxyl group (equation (5)):

$$(5)$$

(XV) (XXIV)

FORMATION OF CHOLESTEROL, ERGOSTEROL AND PHYTOSTEROLS

An excellent review by Goad[77] covers the recent advances in the terminal steps leading to the synthesis of cholesterol in animal tissues, ergosterol in yeasts and fungi and phytosterols in plants and algae. A carrier protein for squalene and the water-insoluble precursors of cholesterol has since been isolated from rat liver by two groups[79,80] and implicated as a vehicle for transferring these substrates across a hydrophilic environment to the sterol synthesizing enzymes. It is felt that a detailed account of the pathways and mechanisms concerned would be out of place in this general text but a brief description will be attempted illustrating some of the salient features.

Lanosterol to Cholesterol

Enzymic conversion of lanosterol into cholesterol requires, without specifying the order in which the modifications are accomplished, removal of the methyl group at C-14 and the gem-dimethyl group at C-4, migration of the nuclear double bond from C-8 to C-5 via the intermediacy of the Δ^7 and $\Delta^{5,7}$ products and reduction of the Δ^{24}-bouble bond in the side-chain (Scheme 7.10). The preferred route, if any, has not been established and it may vary from tissue to tissue. It should be emphasized again at this stage that it is essential for meaningful results in these biosynthetic studies that the intermediates concerned must be rigidly purified to remove all traces of the closely related steroid contaminants, prior to radiochemical assay and subsequent degradation.

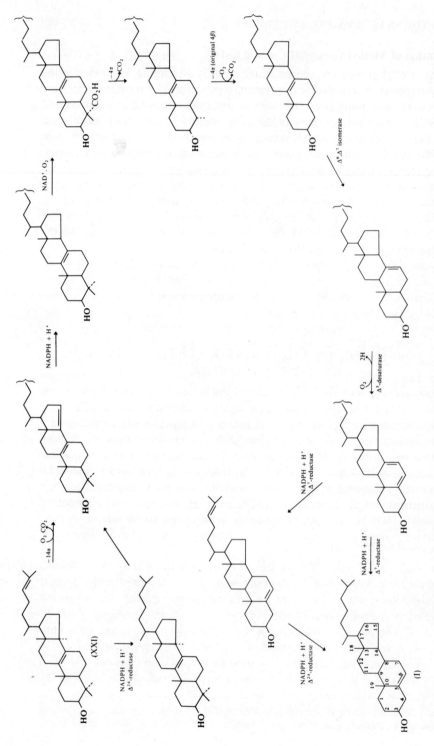

Scheme 7.10 Possible pathways for the conversion of lanosterol (XXI) into cholesterol (I) *via* products possessing Δ^{24}-unsaturation or a saturated side-chain. 3-Ketonic intermediates have been omitted

Oxidation of Methyl Groups at C-4 and C-14

The methyl groups are all eliminated as CO_2 after initial oxidation to the corresponding carboxylic acid (presumably *via* the intermediate formation of alcohol and aldehyde) after attack by a microsomal oxygenase and NADPH. Decarboxylation is NAD^+-dependent.[81] It has been established that the methyl group attached to C-14 is removed first, as 14-methyl- or 4,14-dimethyl sterols have not been detected in animal tissues.[82] Activation of the steroid nucleus necessary for the elimination of the carboxyl group attached to C-14 of the lanosterol derivative may proceed *via* the intermediacy of a $\Delta^{8,14}$-diene since the 15α-tritium atom is lost during the synthesis of cholesterol from $[2\text{-}^3H_2]$mevalonate.[83] Resaturation of the Δ^{14}-bond is effected by NADPH where the hydrogen is directed to the 14α-position, and a proton from the medium.[84]

With regard to the loss of the 4,4′-dimethyl carbon atoms, Clayton's group[85] recently examined this problem with the help of $[2\text{-}^{14}C,5\text{-}^3H]$-mevalonate as tracer substrate in the incubation system. The original 4α-methyl group of lanosterol is derived from C-2 of mevalonate. Squalene, lanosterol (and dihydrolanosterol), the 4α-monomethyl derivative and cholesterol were later isolated and measurements of the $^3H/^{14}C$ ratios of these various intermediates were recorded. It was concluded from the comparison of the values obtained for the sterols with that given by squalene (taken as an internal reference to give a base value) that the 4α-group was attacked first but that the remaining 4β-group underwent epimerization to the 4α-configuration prior to elimination at the next stage. This 4α-product gave a value indicative of loss of a ^{14}C-labelled carbon atom. In confirmation, 4β-methylcholest-7-en-3β-ol-4α-carboxylic acid was identified in aerobic incubation systems deprived of NAD^+, when decarboxylation would be prevented.[81] The original 4α-methyl group was therefore oxidized first.

A further facet of the demethylation process at C-4 rests in the participation of 3-keto intermediates at both stages, formed by oxidation of the 3β-hydroxyl group, in order to facilitate the decarboxylation (Scheme 7.11).

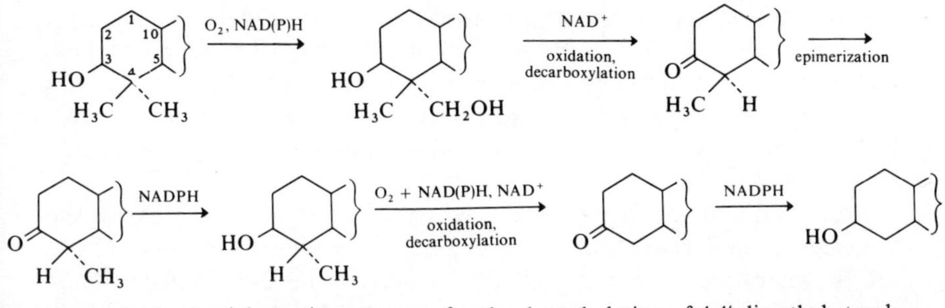

Scheme 7.11   Partial reaction sequence for the demethylation of 4,4′-dimethyl sterols

Restoration of the hydroxyl group is maintained by reduction with a 3-keto-steroid reductase and NADPH. In the absence of NADPH, stoichiometric amounts of ketonic products are observed, compared with loss of CO_2 and consumption of 3β-hydroxy-substrate.

Production of Δ^5-Unsaturation

The conversion of the Δ^8-bond to Δ^5 that appears in cholesterol entails initial isomerization to the Δ^7-position by an anaerobic process. The reactivity of this conversion increases with decreasing degree of methylation.[86] This is followed by the insertion of a double bond at C-5 by an oxygen-dependent process to give the $\Delta^{5,7}$-diene.[87] The mechanism awaits full clarification but is mediated by the direct *cis* elimination of the 5α- and 6α-hydrogen atoms.[88] The final stage in cholesterol biosynthesis is effected by saturation of the Δ^7-bond[89] by the appropriate reductase enzyme and NADPH (Scheme 7.12). Hydrogen is transferred to the 7α- and 8β-positions from [$4S$-^{3}H]NADPH and a proton from the medium respectively.[90]

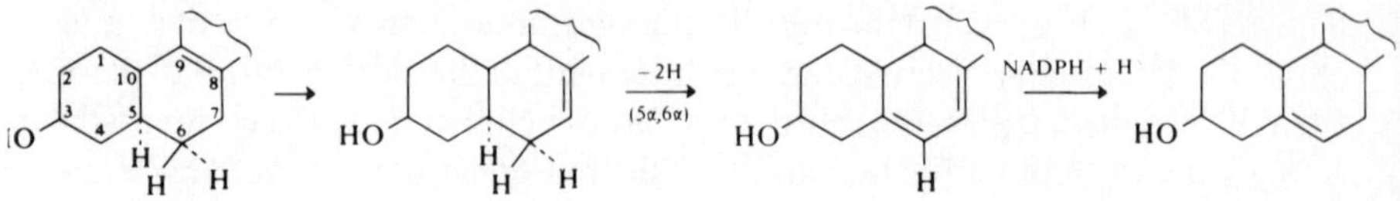

Scheme 7.12   Conversion of the Δ^8-bond into a Δ^5-bond. The reactions illustrated refer to intermediates with either a saturated or Δ^{24}-unsaturated side-chain

Reduction of Δ^{24}-Unsaturation

Two principal divergent pathways for cholesterol synthesis that may be operative in different tissues have been studied; these occur at the level of the Δ^{24}-unsaturated or saturated series (Scheme 7.10). However, saturation of this bond may also take place at the level of many intermediates that may be enzymically converted into cholesterol.[82] The major site of reduction remains unresolved and indeed there may be a great deal of interplay between the possible sterol precursors. Moreover, the Δ^{24}-reductase enzyme is relatively non-specific for sterol substrates.[91] 24,25-Dehydrocholesterol (desmosterol) is saturated by this microsomal enzyme plus NADPH and acts as an immediate precursor of cholesterol, on the basis of various criteria including accumulation *in vivo* after administration of inhibitors such as triparanol.[92]

Examination of the degradation products of cholesterol (after cleavage of the side-chain) derived from rat liver preparations incubated with [2-^{14}C, $4R$-^{3}H$_1$]mevalonate indicates that it possesses the $24R$-configuration.[93] Thus the hydrogen (from the medium) inserted by the reductase enzyme becomes the 24-*pro-S* hydrogen atom. A hydride ion is transferred from NADPH to C-25.[94]

Lanosterol to Ergosterol (II)

Two further problems in relation to ergosterol biosynthesis lie in the introduction of the alkyl group (C-28) and unsaturation at C-22. Investigations into the course of events leading to the transfer of the methyl group were conducted by Lederer and his colleagues[95] using [*Me*-$^2H_3$]methionine. They clearly established that only two deuterium atoms were incorporated into ergosterol (see also Chapter 3 for comments concerning the alkylation of oleate to yield 10-methylstearate). The true methyl donor is S-adenosyl-methionine.[96] It was later observed that the methyl group may be transferred intact to the olefinic double bond in the side-chain[97] and that 24-methylene lanosterol (XXV) proved an effective precursor of ergosterol in yeast cells.[98] However, this trimethylsterol need not be the only substrate *in vivo*.[96] Further experiments in which [24-^3H]- and [26,27-$^{14}C_2$]lanosterol were prepared and incubated with whole cells indicated that alkylation was accompanied by migration of a proton from C-24 to C-25 and loss of one of the original H atoms of S-adenosylmethionine to give a 24-methylene derivative[97] as the first transmethylation intermediate (Scheme 7.13). With respect to insertion of the Δ^{22}-bond, it has been confirmed that a hydrogen is lost from C-23 (and C-22).[99]

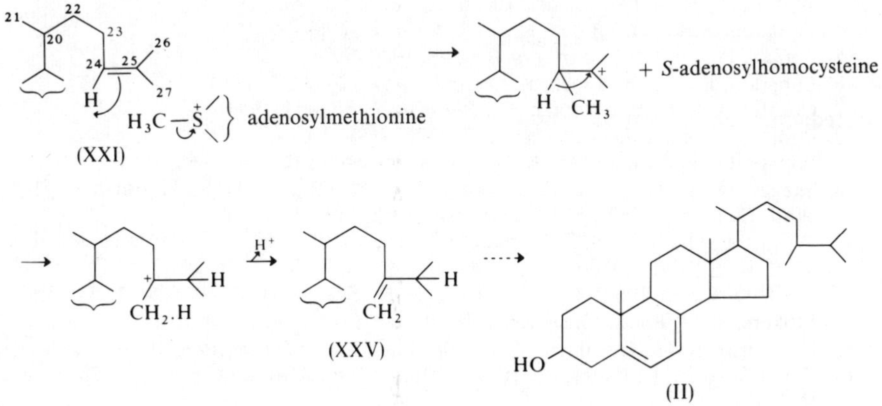

Scheme 7.13   Transfer of methyl group to lanosterol giving rise to 24-methylene lanosterol (XXV) [and subsequently ergosterol (II)]

Cycloartenol to Phytosterols

The intermediary role of cycloartenol in the synthesis of C_{28} and C_{29} sterols in higher plants and algae has been mentioned briefly (p. 166). The remaining features appertaining to these organisms are very similar to those

in the conversion of lanosterol into cholesterol and ergosterol and include the introduction of a methyl or ethyl group at C-24 of the side chain
[campesterol, (XXVI) and β-sitosterol, (XXVII) respectively], with occasional

(XXVI)   (XXVII)

insertion of Δ^{22}-unsaturation. The additional alkyl group is derived from
S-adenosylmethionine after one or two transmethylation reactions to give
24-methylenecycloartanol and 24-ethylidene-products. A further significant
variant from animal systems lies in the ability of plant enzymes to demethylate
initially at C-4. A detailed description of recent work in this field has been
presented by Goad.[77]

REFERENCES

1. Bloch, K., and Rittenberg, D., *J. Biol. Chem.*, **145**, 625 (1942)
2. Rittenberg, D., and Bloch, K., *J. Biol. Chem.*, **160**, 417 (1945)
3. Würsch, J., Huang, R. L., and Bloch, K., *J. Biol. Chem.*, **195**, 439 (1952)
4. Langdon, R. G., and Bloch, K., *J. Biol. Chem.*, **200**, 129 (1953)
5. Popják, G., *Arch. Biochem. Biophys.*, **48**, 102 (1953)
6. Langdon, R. G., and Bloch, K., *J. Biol. Chem.*, **200**, 135 (1953)
7. Cornforth, J. W., and Popják, G., *Biochem. J.*, **58**, 403 (1954)
8. Popják, G., *Annu. Rev. Biochem.*, **27**, 533 (1958)
9. Popják, G., and Cornforth, J. W., *Advan. Enzymol.*, **22**, 281 (1960)
10. Skeggs, M. R., Wright, L. D., Cresson, E. L., MacRae, G. D. E., Hoffman, C. H.,
    Wolf, D. E., and Folkers, K., *J. Bacteriol.*, **72**, 519 (1956)
11. Wright, L. D., Cresson, E. L., Skeggs, H. R., MacRae, G. D. E., Hoffman, C. H.,
    Wolf, D. E., and Folkers, K., *J. Amer. Chem. Soc.*, **78**, 5273 (1956)
12. Wolf, D. E., Hoffman, C. H., Aldrich, P. E., Skeggs, H. R., Wright, L. D., and
    Folkers, K., *J. Amer. Chem. Soc.*, **79**, 1486 (1957)
13. Cornforth, R. H., Cornforth, J. W., and Popják, G., *Tetrahedron*, **18**, 1351 (1962)
14. Tavormina, P. A., Gibbs, M. H., and Huff, J. W., *J. Amer. Chem. Soc.*, **78**, 4498
    (1956)
15. Cornforth, J. W., Cornforth, R. H., Popják, G., and Gore, I. Y., *Biochem. J.*, **69**,
    146 (1958)
16. Tavormina, P. A., and Gibbs, M. H., *J. Amer. Chem. Soc.*, **78**, 6210 (1956)
17. Witting, L. A., Knauss, H. J., and Porter, J. W., *Fed. Proc.*, **18**, 353 (1959)
17a. Brunengraber, H., Sabine, J. R., Boutry, M., and Lowenstein, J. M., *Arch.
    Biochem. Biophys.*, **150**, 392 (1972)
18. Ferguson, J. J., Jr., and Rudney, H., *J. Biol. Chem.*, **234**, 1072, 1076 (1959)
19. Rudney, H. In *The Biosynthesis of Terpenes and Sterols*, p. 75. Ed. by Wolstenholme,
    G. E. W., and O'Connor, M. Churchill Ltd., London, 1959

20. Lynen, F., Henning, U., Bublitz, C., Sorbo, B., and Kröplen-Rueff, L., *Biochem. Z.*, **330**, 269 (1958)
21. Bachhawat, B. K., Robinson, W. G., and Coon, M. J., *J. Biol. Chem.*, **216**, 727 (1955)
22. Stewart, P. R., and Rudney, H., *J. Biol. Chem.*, **241**, 1222 (1966)
23. Rudney, H., Stewart, P. R., Majerus, P. W., and Vagelos, P. R., *J. Biol. Chem.*, **241**, 1226 (1966)
24. Kornblatt, J. A., and Rudney, H., *J. Biol. Chem.*, **246**, 4424 (1971)
25. Corey, E. J., Gregoriou, G. A., and Peterson, D. H., *J. Amer. Chem. Soc.*, **80**, 2338 (1958)
26. Knappe, J., Ringelmann, E., and Lynen, F., *Biochem. Z.*, **332**, 195 (1959)
27. Bucher, N. L. R., Overath, P., and Lynen, F., *Biochim. Biophys. Acta*, **40**, 491 (1960)
28. Cornforth, R. H., Fletcher, K., Hellig, H., and Popják, G., *Nature (London)*, **185**, 923 (1960)
29. Rétey, J., von Stetten, E., Coy, U., and Lynen, F., *Eur. J. Biochem.*, **15**, 72 (1970)
30. Durr, I. F., and Rudney, H., *J. Biol. Chem.*, **235**, 2572 (1960)
31. Siperstein, M. D., and Fagan, V. M., *J. Biol. Chem.*, **241**, 602 (1966)
32. Linn, T. C., *J. Biol. Chem.*, **242**, 990 (1967)
33. White, L. W., and Rudney, H., *Biochemistry*, **9**, 2713 (1970)
34. White, L. W., and Rudney, H., *Biochemistry*, **9**, 2725 (1970)
35. Back, P., Hamprecht, B., and Lynen, F., *Arch. Biochem. Biophys.*, **133**, 11 (1969)
36. Siperstein, M. D., Gyde, A. M., and Morris, H. P., *Proc. Nat. Acad. Sci. U.S.*, **68**, 315 (1971)
37. Lindberg, M., Yuan, C., de Waard, A., and Bloch, K., *Biochemistry*, **1**, 182 (1962)
38. Popják, G., and Cornforth, J. W., *Biochem. J.*, **101**, 553 (1966)
39. Cornforth, J. W., Cornforth, R. H., Popják, G., and Yengoyan, L., *J. Biol. Chem.*, **241**, 3970 (1966)
40. Cornforth, J. W., Cornforth, R. H., Donninger, C., Popják, G., Ryback, G., and Schroepfer, G. J., *Proc. Roy. Soc. Ser. B.*, **163**, 436 (1966)
41. Donninger, C., and Popják, G., *Proc. Roy. Soc. Ser. B*, **163**, 465 (1966)
42. Cornforth, J. W., Cornforth, R. H., Donninger, C., and Popják, G., *Proc. Roy. Soc. Ser. B.*, **163**, 492 (1966)
43. Cornforth, J. W., and Cornforth, R. H. In *Natural Substances Formed Biologically from Mevalonic Acid* (Biochemical Society Symposium, No. 29), p. 5. Ed. By Goodwin, T. W., Academic Press Inc., London and New York, 1970
44. Cahn, R. S., Ingold, C. K., and Prelog, V., *Experientia*, **12**, 81 (1956); *Angew. Chem. Intern. Ed. (England)*, **5**, 385 (1966)
45. Grovenstein, E., and Lee, D. E., *J. Amer. Chem. Soc.*, **75**, 2639 (1953); Cristol, S. J., and Norris, W. P., *J. Amer. Chem. Soc.*, **75**, 2645 (1953)
46. Agranoff, B. W., Eggerer, H., Henning, U., and Lynen, F., *J. Biol. Chem.*, **235**, 326 (1960)
47. Shah, D. H., Cleland, W. W., and Porter, J. W., *J. Biol. Chem.*, **240**, 1946 (1965)
48. Lynen, F., Agranoff, B. W., Eggerer, H., Henning, U., and Möslein, E. M., *Angew. Chem.*, **71**, 657 (1959)
49. Tefft, R. E., Goodwin, T. W., and Simpson, K. L., *Biochem. J.*, **117**, 921 (1970)
50. Benedict, C. R., Kett, J., and Porter, J. W., *Arch. Biochem. Biophys.*, **110**, 611 (1965)
51. Holloway, P. W., and Popják, G., *Biochem. J.*, **104**, 57 (1967)
52. Johnson, W. S., and Bell, R. A., *Tetrahedron Lett.*, **No. 12**, 27 (1960)

53. Popják, G., Goodman, D. S., Cornforth, J. W., Cornforth, R. H., and Ryhage, R., *J. Biol. Chem.*, **236**, 1934 (1961)
54. Childs, C. R., and Bloch, K., *J. Biol. Chem.*, **237**, 62 (1962)
55. Popják, G., Cornforth, J. W., Cornforth, R. H., Ryhage, R., and Goodman, D. S., *J. Biol. Chem.*, **237**, 56 (1962)
56. Levy, H. P., Talalay, P., and Vennesland, B. In *Progress in Stereochemistry*, Vol. 3, p. 299. Butterworth & Co. Ltd., London, 1952
57. Gawron, O., and Fondy, T. P., *J. Amer. Chem. Soc.*, **81**, 6333 (1959); Anet, F. A. L., *J. Amer. Chem. Soc.*, **82**, 994 (1960)
58. Popják, G. In *Natural Substances Formed Biologically from Mevalonic Acid* (Biochemical Society Symposium No. 29), p. 17. Ed. by Goodwin, T. W. Academic Press Inc., London and New York, 1970
59. Rilling, H. C., *J. Biol. Chem.*, **241**, 3233 (1966)
60. Epstein, W. W., and Rilling, H. C., *J. Biol. Chem.*, **245**, 4597 (1970)
61. Edmond, J., Popják, G., Wong, S., and Williams, V. P., *J. Biol. Chem.*, **246**, 6254 (1971)
62. Sofer, S. S., and Rilling, H. C., *J. Lipid Res.*, **10**, 1835 (1969)
63. Rilling, H. C., *J. Lipid Res.*, **11**, 480 (1970)
64. Clayton, R. B., *Quart. Rev. Chem. Soc.*, **19**, 168 (1965)
65. Ruzicka, L., *Experientia*, **9**, 357 (1953)
66. Eschenmoser, A., Ruzicka, L., Jeger, O., and Arigoni, D., *Helv. Chim. Acta*, **38**, 1890 (1955)
67. Tchen, T. T., and Bloch, K., *J. Biol. Chem.*, **226**, 921, 931 (1957)
68. Maudgal, R. K., Tchen, T. T., and Bloch, K., *J. Amer. Chem. Soc.*, **80**, 2589 (1958)
69. Cornforth, J. W., Cornforth, R. H., Pelter, A., Horning, M. G., and Popják, G., *Tetrahedron*, **5**, 311 (1959)
70. Willett, J. D., Sharpless, K. B., Lord, K. E., van Tamelen, E. E., and Clayton, R. B., *J. Biol. Chem.*, **242**, 4182 (1967)
71. Corey, E. J., Russey, W. E., and Ortiz de Montellano, P. R., *J. Amer. Chem. Soc.*, **88**, 4750 (1966)
72. Corey, E. J., and Russey, W. E., *J. Amer. Chem. Soc.*, **88**, 4751 (1966)
73. van Tamelen, E. E., and Curphey, T. J., *Tetrahedron Lett.*, **3**, 121 (1962)
74. Yamamoto, S., Lin, K., and Bloch, K., *Proc. Nat. Acad. Sci. U.S.*, **63**, 110 (1969)
75. Shechter, I., Sweat, F. W., and Bloch, K., *Biochim. Biophys. Acta*, **220**, 463 (1970)
76. Yamamoto, S., and Bloch, K., *J. Biol. Chem.*, **245**, 1670 (1970)
77. Goad, L. J. In *Natural Substances Formed Biologically from Mevalonic Acid* (Biochemical Society Symposium No. 29), p. 45. Ed. by Goodwin, T. W. Academic Press Inc., London and New York, 1970
78. Zander, J. M., Greig, J. B., and Caspi, E., *J. Biol. Chem.*, **245**, 1247 (1970)
79. Scallen, T. J., Schuster, M. W., and Dhar, A. K., *J. Biol. Chem.*, **246**, 224 (1971)
80. Ritter, M. C., and Dempsey, M. E., *J. Biol. Chem.*, **246**, 1536 (1971)
81. Miller, W. L., and Gaylor, J. L., *J. Biol. Chem.*, **245**, 5369, 5375 (1970)
82. Frantz, I. D., and Schroepfer, G. J., *Annu. Rev. Biochem.*, **36**, 691 (1967)
83. Akhtar, M., Watkinson, I. A., Rahimtula, A. D., Wilton, D. C., and Munday, K. A., *Biochem. J.*, **111**, 757 (1969)
84. Akhtar, M., Rahimtula, A. D., Watkinson, I. A., Wilton, D. C., and Munday, K. A., *J. Chem. Soc. D* (*Chem. Commun.*), p. 149 (1969)
85. Rahman, R., Sharpless, K. B., Spencer, T. A., and Clayton, R. B., *J. Biol. Chem.*, **245**, 2667 (1970)
86. Gaylor, J. L., Delwiche, C. V., and Swindell, A. C., *Steroids*, **8**, 353 (1966)

87. Dewhurst, S. M., and Akhtar, M., *Biochem. J.*, **105,** 1187 (1967)
88. Paliokas, A. M., and Schroepfer, G. J., *J. Biol. Chem.*, **243,** 453 (1968)
89. Dempsey, M. E., *J. Biol. Chem.*, **240,** 4176 (1965)
90. Wilton, D. C., Munday, K. A., Skinner, S. J. M., and Akhtar, M., *Biochem. J.*, **106,** 803 (1968)
91. Avigan, J., Goodman, D. S., and Steinberg, D., *J. Biol. Chem.*, **238,** 1283 (1963)
92. Avigan, J., Steinberg, D., Vroman, H. E., Thompson, M. J., and Mosettig, E., *J. Biol. Chem.*, **235,** 3123 (1960)
93. Greig, J. B., Varma, K. R., and Caspi, E., *J. Amer. Chem. Soc.*, **93,** 760 (1971)
94. Akhtar, M., Munday, K. A., Rahimtula, A. D., Watkinson, I. A., and Wilton, D. C., *J. Chem. Soc. D (Chem. Commun.)*, p. 1287 (1969)
95. Jauréguiberry, G., Law, J. H., McCloskey, J. A., and Lederer, E., *Biochemistry*, **4,** 347 (1965)
96. Katsuki, H., and Bloch, K., *J. Biol. Chem.*, **242,** 222 (1967)
97. Akhtar, M., Hunt, P. F., and Parvez, M. A., *Biochem. J.*, **103,** 616 (1967)
98. Akhtar, M., Parvez, M. A., and Hunt, P. F., *Biochem. J.*, **100,** 38c (1966)
99. Akhtar, M., Parvez, M. A., and Hunt, P. F., *Biochem. J.*, **106,** 623 (1968)

CHAPTER 8
Biosynthesis of Isoprenoid Compounds Derived from Geranylgeranyl Pyrophosphate

The biosynthesis of squalene and sterols from farnesyl pyrophosphate (farnesyl-PP) has been dealt with in the previous chapter. The remaining metabolites formed from the diterpene (C_{20}) geranylgeranyl-PP (I) and higher isoprenologues will now be considered and their mode of biosynthesis is briefly outlined in Scheme 8.1. A vast array of cyclic terpene products

Isopentenyl-PP ← Mevalonate ← 3 Acetyl-CoA
$\downarrow C_5$

Geranyl-PP (C_{10}) → Monoterpenes
$\downarrow C_5$

Farnesyl-PP (C_{15}) → Sesquiterpenes

C_{15}

C_5 → Squalene (C_{30}) → Triterpenes (sterols; tetrahymanol)

Geranylgeranyl-PP (C_{20}) → Diterpenes (gibberellins; chlorophyll, vitamins E and K_1)

C_{20}

$(C_5)_n$ → Phytoene (C_{40}) → Tetraterpenes (carotenes, xanthophylls)

Polyisoprenyl-PP → Polyterpenes (polyprenols, rubber, gutta-percha; side-chain of ubiquinone, plastoquinone and vitamin K_2)

Scheme 8.1   Derivation of isoprenoid products in Nature

including alkaloid derivatives has been characterized. Rather than catalogue the individual idiosyncrasies of the folding patterns achieved during the formation of these compounds, a representative of a diterpene family with biological importance as plant hormones, the gibberellins, has been selected for short discussion below. Phytol (II) is formed by partial saturation of geranylgeranyl-PP and exists widely in Nature as an ester moiety in chlorophyll (the universal green pigment associated with photosynthetic tissue) and as the side-chain of vitamin K_1.

176

(I)

(II)

Tetraterpenes are C_{40} products derived from two molecules of geranyl-geranyl-PP in a manner that is somewhat similar to that encountered in squalene formation. Additional unsaturation, however, is inserted at the centre of the resulting molecule to yield phytoene. Carotenes and the related oxygenated xanthophylls form an extensive group of natural pigments that belong to this class and are termed carotenoids. They are widely distributed among the plant kingdom where they are concentrated in the grana of chloroplasts within the leaves of all green plants and algae. In addition, they are commonly found in fruits, flowers (mainly as xanthophylls), many fungi and certain bacteria. Although carotenoids may be located in animal tissues they are not synthesized by higher animals. These red and yellow products are soluble in lipids (and organic solvents) and exert a photo-receptor role in photosynthetic tissues by virtue of their strong ultraviolet- and visible light-absorption properties. They may also function by protecting these systems against possible damage caused by photo-oxidation.

Further elongation of geranylgeranyl-PP by repeated condensation with isopentenyl-PP gives rise to linear long-chain products. Rubber, a high molecular weight polymer composed of isoprene residues, and the poly-prenols which contain up to 120 carbon atoms that are found in diverse tissues and organisms are examples of such metabolites. These latter com-pounds exert a carrier role in bacteria (and probably elsewhere) during the synthesis of insoluble polymers that are laid down outside the cell-membrane. Somewhat shorter-chain residues are present in the prenyl side-chains attached to aromatic systems in products that play important roles in electron transport and other essential metabolic activities. The function of the terpenoid quinones and tocopherols resides in the aromatic portion of the molecule but the side-chains confer lipid solubility (irrespective of chain-length or degree of unsaturation) on these products that enables them to be positioned in membranous structures within the cell at sites related to their molecular mode of action. The nature of the polyprenyl (derived) side-chain may be important in determining the pattern of absorption and distribution amongst various tissues.

BIOSYNTHESIS OF DITERPENES

Gibberellins

These metabolites have been identified in tissues of higher plants and are important promoters of plant growth and differentiation.[1] Knowledge of their biosynthetic pathway may indicate possible sites of control and thereby

assist in the regulation of these processes. Early studies[2] revealed that gibberellic acid (GA$_3$, (III)), one of the most important hormones in this series,
was derived from mevalonate and suggested that a biosynthetic relationship
with other diterpenes might exist. From structural considerations and an
understanding of the mechanism established for the cyclization of squalene,
it was proposed that synthesis of the gibberellin skeleton might proceed
via a proton-initiated attack on all-*trans*-geranylgeranyl-PP (I).[3,4] This
would be followed by a concerted electron rearrangement and loss of a
proton to give the stabilized bicyclic product (IV) which in turn is transformed
to the tetracyclic hydrocarbon, kaurene (V), by means of a secondary cyclization (Scheme 8.2).

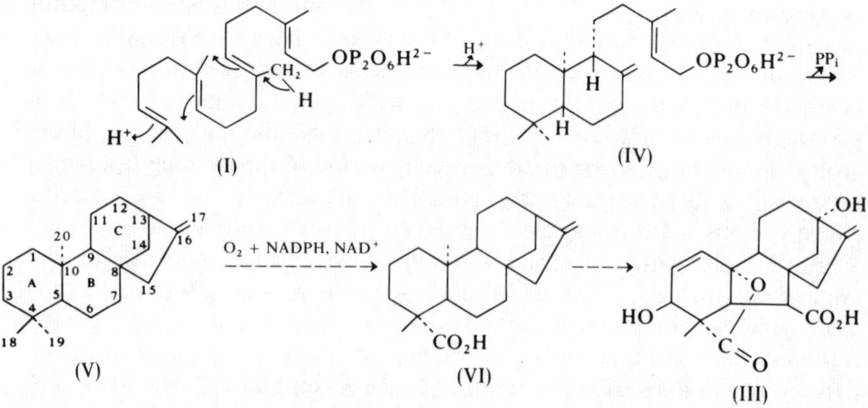

Scheme 8.2   Partial reaction sequence for the conversion of geranylgeranyl-PP into
gibberellic acid (III). PP$_i$ represents HP$_2$O$_7^{3-}$

The reaction responsible for the first cyclization into copalyl-PP (IV)
has since been examined at the enzyme level with a preparation obtained
from the fungus *Gibberella fujikuroi* (the perfect stage of the organism identified as *Fusarium moniliforme*).[5] Moreover, the ability of similar extracts of
this fungus and higher plants to convert this substrate into kaurene was also
demonstrated. Cyclase activity associated with both conversions could not
be resolved in the fungal synthetase and presumably resides in the same
enzyme.[6]

The precursor activity of kaurene[1,7] and its alcohol, aldehyde and acid
(VI)[8] derivatives for gibberellic acid formation in fungi had previously been
established; in the course of this reaction sequence ring B undergoes con-
traction to a cyclopentanoic acid. Oxidation of the 4α-methyl group (C-19)
proceeds in the microsomes *via* an oxygen- and NADPH-dependent mechanism, analogous to that determined for cholesterol synthesis, but the carboxyl
group is retained in the product as the lactone. Again, the methyl groups
display their individuality as only the 4α-carbon is oxidized.

Phytol (II) and Chlorophyll

One of the most widely-occurring of all natural products is chlorophyll which is invariably found as the green pigment in photosynthetic tissues. Information leading to a knowledge of chlorophyll biosynthesis including the stage at which phytol is incorporated has been obtained mainly from studies with algae and photosynthetic bacteria. Indeed, it was first established with Chlorella *mutants* that the reactions responsible for the formation of chlorophyll and haem followed the same pathway as far as protoporphyrin.[9] Conversion of this tetrapyrrole into chlorophyll *a* requires firstly, insertion of Mg^{2+} followed by, *inter alia*, reduction of a vinyl side-chain, methylation, oxidation and cyclization of a propionic acid unit and finally, esterification of the remaining propionic acid residue attached to pyrrole ring D with phytol.[10] It is this latter grouping that confers the lipid solubility and biological activity on chlorophyll. Chlorophyll *b* is derived by conversion of the methyl group on ring B into a formyl group. Esterification proceeds by reaction of phytol with chlorophyllide *a* (containing the free carboxyl group) and is mediated by chlorophyllase (chlorophyll: chlorophyllide hydrolase, EC 3.1.1.14), an enzyme that has long been known to act hydro-lytically.[11] No evidence of a requirement for phytyl-PP has been found.[12]

The stereochemistry of the residual α,β-double bond in phytol and the saturated groups formed after reduction has been examined; the isoprenoid units are all biogenetically *trans*.[13]

BIOSYNTHESIS OF TETRATERPENES

Carotenes

These are derived from the colourless phytoene (VII) after initial condensation of two molecules of geranylgeranyl-PP by a 'tail to tail' mechanism with elimination of pyrophosphate. Alkylation to give phytoene may occur after isomerization of one C_{20} substrate into geranyllinaloyl-PP (Ia),[14] but the oxidation state remains unchanged in this system unlike that in the conversion of farnesyl-PP into squalene. This possibility is expressed in equation (1):

$$\text{R}-\text{CH}_2-\overset{\overset{\displaystyle \text{CH}_3}{|}}{\text{C}}=\text{CH}-\text{CH}_2 \quad \overset{\overset{\displaystyle \text{OPP}}{|}}{\text{C}} \quad \text{CH}_2=\text{CH}-\overset{\overset{\displaystyle \text{OPP}}{|}}{\underset{\underset{\displaystyle \text{CH}_3}{|}}{\text{C}}}-\text{CH}_2-\text{R} \xrightarrow{\text{2PP}}$$

(I)   (Ia)   (1)

$$\text{R}-\text{CH}_2-\overset{\overset{\displaystyle \text{CH}_3}{|}}{\text{C}}=\text{CH}-\text{CH}=\text{CH}-\text{CH}=\overset{\underset{\underset{\displaystyle \text{CH}_3}{|}}{}}{\text{C}}-\text{CH}_2-\text{R}$$

(VII)

Tetraterpenes (C_{40}) do not undergo such extensive cyclization as the di- or triterpene groups since the additional central double bond in phytoene gives a conjugated triene structure that prevents similar folding patterns. The lack of these products in Nature gives added significance to the evidence that phytoene, rather than the squalene analogue lycopersene, is the initial hydrocarbon formed. The linear chain is susceptible to step-wise dehydrogenation giving rise to substances with a conjugated system of double bonds and hence pronounced ultraviolet- and visible light-absorption spectra. The limited ability to form cyclic products is expressed late in the biosynthetic sequence and results in the presence of terminal monocyclic or bicyclic groups.

Degradative studies first established that the hydrocarbon β-carotene (VIII) was indeed derived from acetate *via* the intermediacy of mevalonate in fungi and the phytoflagellate *Euglena gracilis* (Scheme 8.3).[15-17] Further

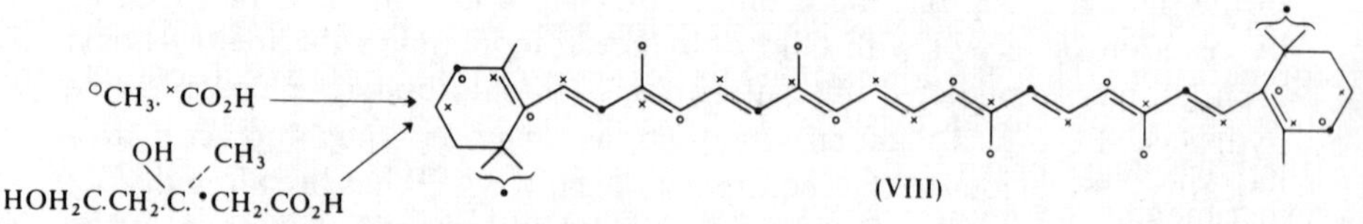

°CH_3.ˣCO_2H ⟶

$\underset{HOH_2C.CH_2.C.ˣCH_2.CO_2H}{\overset{OH \quad CH_3}{\big|}}$

(VIII)

Scheme 8.3   Distribution of ^{14}C-labelled atoms in β-carotene (VIII) from $[1\text{-}^{14}C]$-acetate (ˣ), $[2\text{-}^{14}C]$acetate (°) and $[2\text{-}^{14}C]$mevalonate (•)

investigations into the mode of synthesis of isoprenoid-derived substances in developing seedlings brought to light the fact that chloroplast membranes were inaccessible to exogenously supplied mevalonate. $^{14}CO_2$, on the other hand, was rapidly fixed into β-carotene and other plastidic components, e.g. phytol, but incorporation into sterols was considerably less efficient.[18] The fixation of CO_2 was dependent on the presence of β-methylcrotonyl-CoA (formed during leucine catabolism) and yielded β-methylglutaconyl-CoA and HMG-CoA.[19] These situations may help to regulate synthesis of various isoprenoid derivatives within the intact cell, by means of compartmentalization of the relevant enzymes into chloroplastidic and extra-plastidic pathways. Goodwin[18] has pointed out that enzymes related to squalene and sterol synthesis (that is including prenyltransferases responsible for the formation of the C_{15} farnesyl residue) reside in the microsomal and cytosol fractions while those required for carotenoid, chlorophyll and terpenoid quinone synthesis (but not ubiquinone formation which is mitochondrial) are confined to the chloroplast. Transferases of rather different specificity are needed for these processes. Phytosterols are transferred to the membranes of developing plastids after illumination of etiolated seedlings, which form minimal quantities of carotenoids and other products essential for photosynthesis. Once formed, however, the chloroplasts synthesize mevalonate from metabolites derived from CO_2 after fixation and hence their complement of isoprenoid compounds.

A further understanding of the mechanism involved in carotenoid formation was gained after treatment of chloroplasts by a process of homogenization in solvent, followed by ultrasonication.[20] These soluble preparations were capable of converting mevalonate into phytoene, the precursor of coloured carotenoids and, in addition, the phytol side-chain of chlorophyll. The conversion into phytoene was independent of light and nucleotide cofactors.

Attention was next focussed on the stereospecific aspects of carotene biosynthesis in plant and fungal tissues by Goodwin and associates who examined them with the aid of doubly-labelled species of mevalonate. Formation of phytoene in carrot root slices occurred with complete retention of the 4R-tritium atoms from [2-^{14}C, 4R-^{3}H$_1$]mevalonate; ^{14}C/^{3}H atomic ratios in the product remained as 8:8 (normalized ratio in substrate was 1:1).[21] The same ratio was observed for squalene synthesized by these preparations. Thus the double bond present in each residue derived from C-4 (and C-3) of mevalonate existed in the *trans* configuration. Details concerning the later stages of carotene synthesis were also clarified by the use of this experimental technique. β-Carotene (VIII) possessed a lower ^{14}C/^{3}H atomic ratio with a value of 8:6, indicative of loss of two tritium atoms from the C-6 and C-6′ positions in the formation of the β-ionone rings (Scheme 8.4);

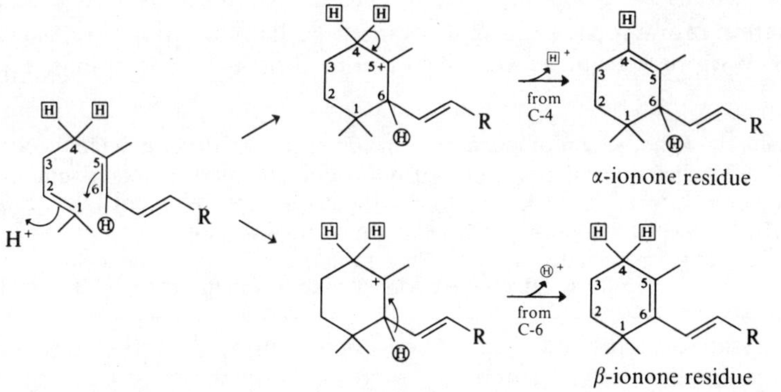

Scheme 8.4 Formation of α- and β-ionone rings *via* a common carbonium ion derived from [4R-^{3}H$_1$]mevalonate (⊕) and [2-^{3}H$_2$]mevalonate (H) (Williams *et al.*, 1967)[22]

these atoms which arose from the 4R position of mevalonate were still present in geranylgeranyl-PP (not isolated) and phytoene. Moreover, α-carotene (IX)

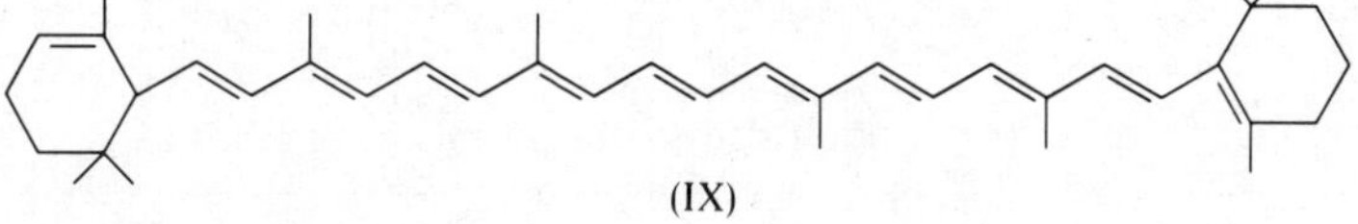

(IX)

which contains a terminal α- and β-ionone ring had a smaller decreased ratio
(8 : 7) and an additional labelled atom was therefore present at C-6 (numbering
system based on carotene). Thus this product could not have arisen from β-
carotene by isomerization. Further observations[22] from work with [2-^{14}C,
2-[3]H$_2$]mevalonate confirmed by application of a similar argument that β-
carotene was not formed from α-carotene. Enzyme assembly of α- and β-
ionone rings within the cyclic carotenoid series occurs independently by
stereospecific elimination of different protons from a common carbonium
ion intermediate (as a consequence of specific alignment at the appropriate
active sites), after a primary proton attack at C-2 of the acyclic precursor
(Scheme 8.4).

The central double bond in phytoene originates from C-5 of two molecules
of mevalonate (C-1 of geranylgeranyl-PP) after removal of two hydrogen
atoms, one from each substrate. Moreover, each subsequent dehydrogenation
leading to the more unsaturated products removes a further hydrogen atom
from a carbon atom that was formerly C-5 of mevalonate. The problem of
detecting whether these reactions were stereospecific and also the mechanism
of condensation were studied with [2-^{14}C, 5R-[3]H$_1$]- and [2-^{14}C, 5-[3]H$_2$]-
mevalonate in tomato fruit (incorporation into lycopene, (XII)) and bean
leaves (incorporation into α- and β-carotene).[14] The values for the ^{14}C/[3]H
atomic ratios obtained in these experiments are outlined in Table 8.1 and are
based on figures determined for squalene as reference compound. The 5R-
tritium atoms were retained and therefore two 5S-atoms were lost in the
formation of phytoene (to give a resultant *cis* configuration[23] at the central
15,15′ position), whereas four 5R-tritium atoms were eliminated in the

Table 8.1   Comparison of incorporation of ^{14}C and ^{3}H from different species of
doubly-labelled mevalonate into various isoprenoid-derived compounds after
incubation with tomato slices and bean leaf slices (from Williams *et al.*, 1967)[14]
The ^{14}C/^{3}H atomic ratios are based on a normalized value of 6 : 11 or 6 : 6 for the
squalene that was isolated and used as an internal reference. The average values
gained from different experiments are recorded in this table

Labelled substrate	Substance isolated		[14]C/^{3}H atomic ratio
[2-^{14}C,5-[3]H$_2$]Mevalonate	Squalene	(C$_{30}$)	6:11
	Phytoene	(C$_{40}$)	8:14·31
	Lycopene	(C$_{40}$)	8:9·94
	α-Carotene	(C$_{40}$)	8:10·06
	β-Carotene	(C$_{40}$)	8:10·33
[2-^{14}C,5R-5-[3]H$_1$]Mevalonate	Squalene		6:6
	Phytoene		8:8·23
	Lycopene		8:3·81
	α-Carotene		8:4·18
	β-Carotene		8:4·15

dehydrogenation of phytoene to lycopene or the cyclic carotenes, presumably one at each each stage of the sequence. The desaturation reactions that give rise to the additional double bonds are therefore all stereospecifically controlled. The individual carotenes were resolved from each other after preliminary treatment and rigorously purified by thin-layer chromatography. A mechanism for the condensation reaction between geranyl-geranyl-PP and the isomeric geranyl linaloyl-PP involved in the formation of phytoene by means of a cis-1,4-elimination has been presented[14] (Scheme 8.5). The transitory positively-charged C_{40} intermediate that forms on loss

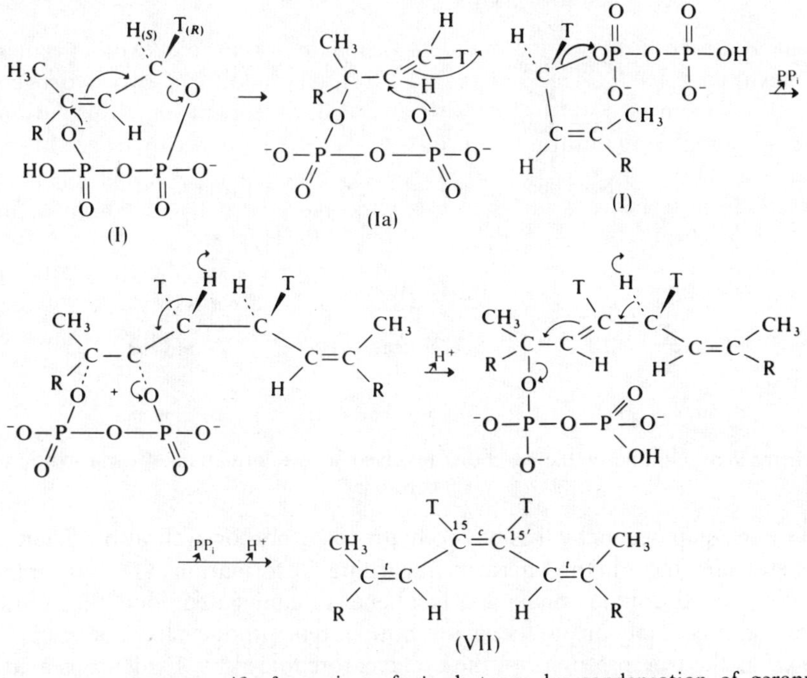

Scheme 8.5 Stereospecific formation of cis-phytoene by condensation of geranyl-geranyl-PP with geranyllinaloyl-PP, derived from [$5R$-$^3H_1$]mevalonate (from Williams et al., 1967)[14]

of pyrophosphate is stabilized in this sequence by loss of a proton to generate the central double bond rather than by gain of a hydride ion from NADPH (which results in the saturated centre in squalene).

Evidence for the early steps in carotenogenesis was based on inhibition studies with diphenylamine and incorporation of precursors into more unsaturated products by chloroplasts or tomato fruit plastids. More directly it was shown[24] that a partially purified preparation from tomato plastids containing prenyl transferases and phytoene synthetase converted isopentenyl-PP (with or without additional farnesyl-PP) or geranylgeranyl-PP into phytoene.[25] Nicotinamide nucleotides and oxygen were not required for these reactions,[25] supporting the mechanism assigned in Scheme 8.5. An improved enzyme system yielded phytoene, its immediate dehydrogenated product phytofluene (5 conjugated double bonds), neurosporene (X) and lycopene (XII).[26]

On the basis of these and other findings, sequences have been proposed for the synthesis of α- and β-carotene and are illustrated in Schemes 8.6 and

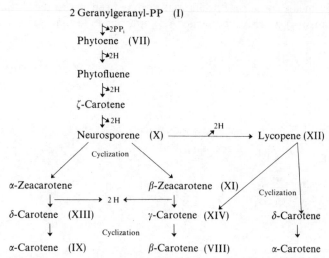

Scheme 8.6   Outline of the reactions involved in the formation of α-carotene and β-carotene.

8.7. Each step occurs by loss of two hydrogen atoms or cyclization. There is general agreement in the Literature regarding the formation of neurosporene (9 conjugated double bonds) and lycopene (11 conjugated bonds) from isoprenoid precursors or phytoene, but both acyclic products have been implicated as the true precursor of the cyclic carotenoids.[18,27] Evidence is available in support of the case that neurosporene exists at a branch-point of carotenoid biosynthesis and that it acts as a precursor for both α- and β-carotene, with the partially cyclized α- and β-zeacarotene (XI) as intermediates. These latter metabolites have been identified in trace amounts in carotenogenic systems.

Contrasting results, however, were first reported from Porter's Laboratory[28] that favoured lycopene (XII) as the immediate precursor of the cyclic carotenoids in higher plant plastids. [15,15'-³H]Lycopene was converted into α- and β-carotene via δ-carotene (XIII) and γ-carotene (XIV) intermediates by spinach chloroplasts and tomato plastids, indicating that cyclization took place after the level of neurosporene.[29] α- and β-Zeacarotene would therefore behave as secondary end-products. NADP⁺ was required

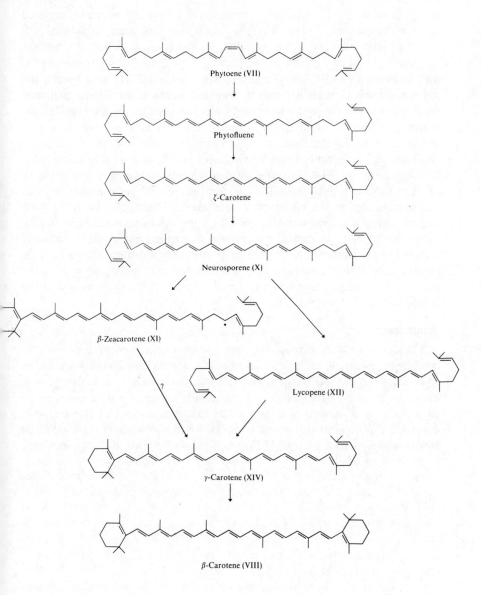

Scheme 8.7   Reactions leading to the formation of β-carotene. Sites of desaturation have been marked (•)

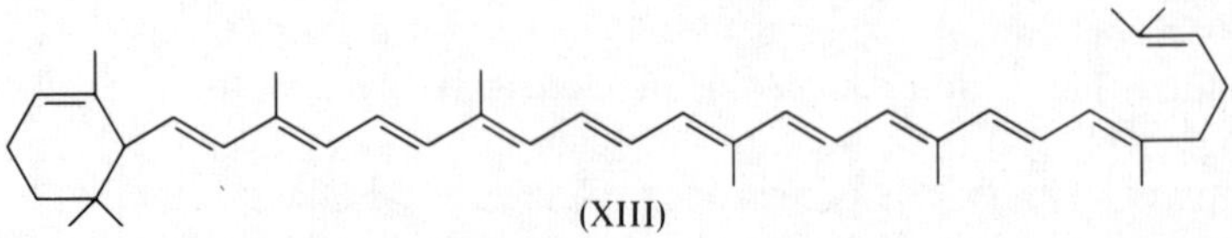

(XIII)

for maximum activity with these preparations but FAD was specifically needed for the formation of β-carotene in soluble extracts. The products were purified by column chromatography, crystallization and gas–liquid-chromatography of the perhydro-derivatives. Attempts were not made in this work to investigate the precursor activity of neurosporene under the same conditions but it is difficult to obtain a good source of this material for labelling purposes. More recently, the use of inhibitors has been applied in an attempt to resolve this question.[30] Incubation of a *Mycobacterium* sp. with nicotine resulted in the inhibition of formation of the cyclic β-carotene and accumulation of lycopene. After washing, the cells recovered their ability to synthesize the cyclic carotenoids (mainly β-carotene with some γ-carotene), but not β-zeacarotene, at the expense of the preformed lycopene. This product, therefore, appears to be the metabolite that undergoes cyclization rather than neurosporene, in accordance with the pathway expressed at the right-hand side of Scheme 8.6.

Xanthophylls

These oxygenated carotenoids are found together with carotenes in photosynthetic tissue within the chloroplasts. The oxygen function is inserted at a late stage in their biosynthesis and may be present as a hydroxy, oxo, epoxy or methoxy group with a characteristic pattern for different types of organisms. It is derived from molecular O_2 as in lutein (3,3′-dihydroxy-α-carotene, (XV))[31] and violaxanthin (XVI),[31] and also antheraxanthin (XVII) (epoxy and hydroxyl groups).[32] Following their detailed investigations into

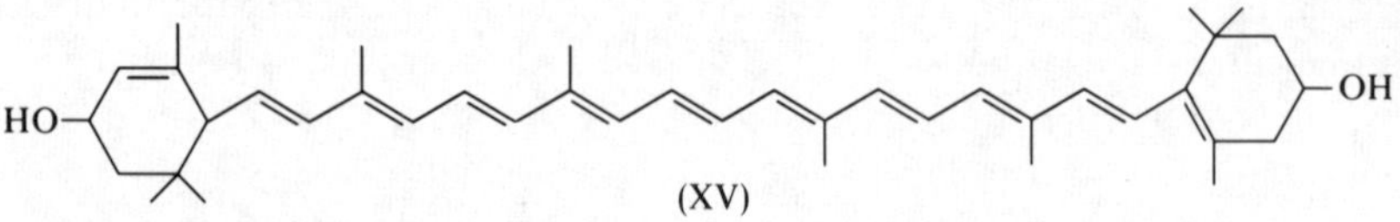

(XV)

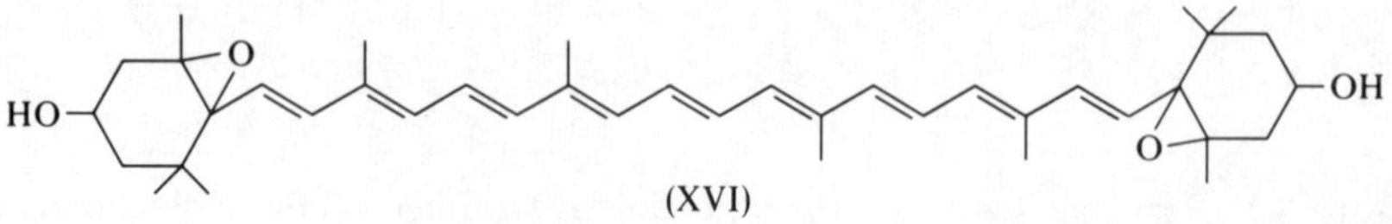

(XVI)

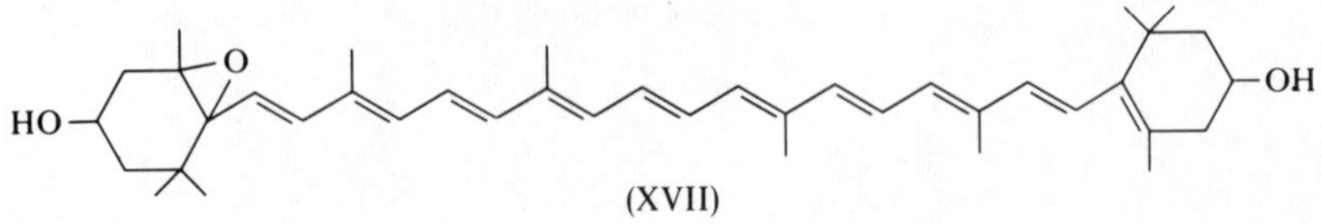

(XVII)

the stereochemical aspects of carotene synthesis, Goodwin's group[33] extended their studies to problems concerned with xanthophyll formation in higher plants, in particular, that relating to the insertion of the hydroxyl group at C-3 and C-3′ of the cyclic carotenes (e.g. in the production of lutein). These carbon atoms derive from C-5 of mevalonate and both hydrogen atoms are retained in these positions in the formation of carotenes.[14] Accordingly, mevalonate that was asymmetrically labelled with ^{3}H at the prochiral C-5 atom proved most suitable for this purpose. Hydroxylation occurred with loss of the 5R-tritium atom.[33]

Retinal (Retinene, Vitamin A Aldehyde) (XVIII)

Retinal and the alcohol derivative have important functions in reproductive processes and in the maintenance of epithelial tissue while the former plays an essential role in vision as a protein-bound component in visual pigments. A β-carotene 15,15′-oxygenase has been isolated from rat intestinal mucosa[34,35] and liver[36] that cleaves β-carotene (provitamin A$_1$) at the central double bond into two fragments of all-*trans*-retinal by reaction with molecular O$_2$ (equation (2)):

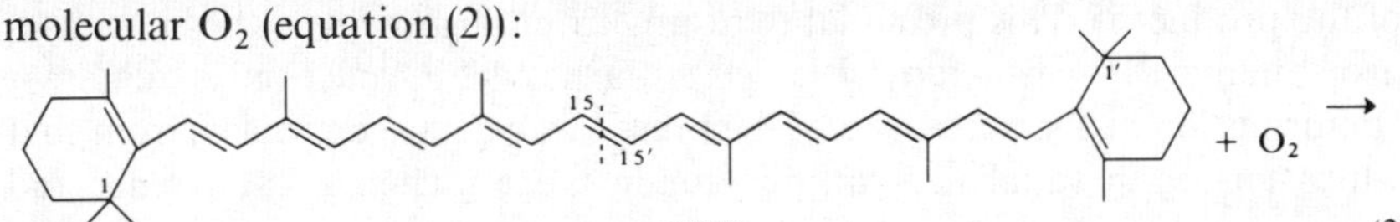

(VIII)   + O$_2$ →

(2)

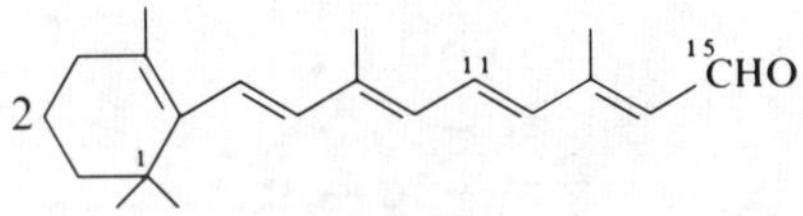

(XVIII)

The mechanism is typical of that catalysed by a dioxygenase enzyme since nicotinamide nucleotides are not involved and the hydrogen atoms attached to the central C-15 and C-15′ atoms of β-carotene are both retained in retinal formed by mucosal homogenates.[34] This product is reduced *in vivo* by an NADH-mediated enzyme to retinol (vitamin A) which is esterified with long-chain fatty acids and transported as chylomicrons from intestinal mucosa to the liver for storage *via* the lymphatic system. It is eventually remobilized as the free alcohol prior to removal to the eye and other tissues. In the eye, the retinol is reoxidized to all-*trans*-retinal and converted to the Δ^{11}-*cis*-stereoisomer (XVIIIa) which combines in the dark with the protein opsin to give the visual pigment rhodopsin.[36a]

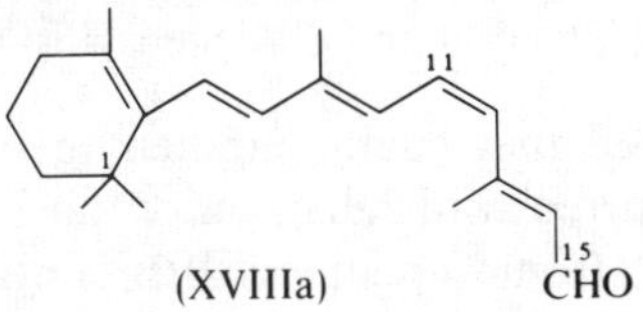

(XVIIIa)   CHO

BIOSYNTHESIS OF POLYTERPENES

Discussion has so far ranged over the assembly of products derived from dimethylallyl-PP with up to three molecules of isopentenyl-PP but a number of systems exist that are responsible for further successive reactions of geranylgeranyl-PP with isopentenyl-PP by repeated condensations. These will now be described.

Rubber

This extremely useful natural product has a very high molecular weight with the structure $H[C_5H_8]_n$-PP in which n varies from 500–5000. It is formed in the latex of many genera of Angiosperms but in especially good yield in *Hevea brasiliensis*, where it may readily be obtained by cutting into the bark. Its biosynthesis proceeds by chain extension at the surface of existing rubber particles where the final condensation with isopentenyl-PP takes place.[37] The reactions engaged in are similar to those concerned with the formation of farnesyl-PP but the stereospecific elimination of the protons from C-2 of the substrates give rise to a *cis* orientation in the double bonds of the product.[38] This presumably results from alignment of the substrates (isopentenyl-PP, *cis*-geranyl-PP, *cis,cis*-farnesyl-PP, etc.) in a different conformation at the active site of the prenyl transferase enzyme(s) from that which ensues in squalene synthesis, for instance (Scheme 8.8). Archer and

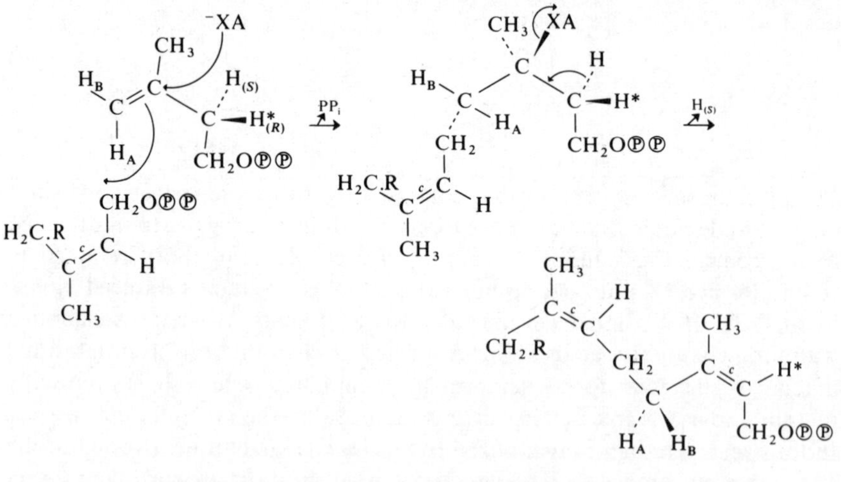

Scheme 8.8 Stereospecificities involved in the transferase reactions leading to the formation of *cis*-polyprenols (Cornforth *et al.*, 1966).[38] $R = (CH_3)_2.C:CH.CH_2.$ $[CH_2.C(CH_3):CH.CH_2]_n—$; ⓟⓟ and PP_i denote $HP_2O_6^{2-}$ and $HP_2O_7^{3-}$ respectively

colleagues[39] observed that chain-lengthening of *cis*-1,4-polyisoprenyl-PP, used as primer, occurred with retention of the 4S-tritium atom from [4S-4-$^3H_1$]mevalonate (2R- from isopentenyl-PP), contrasting with retention of the

4*pro-R* hydrogen in the formation of *trans*-geranyl-PP and *trans,trans*-farnesyl-PP.

The isomeric all-*trans*-gutta percha[40] is produced by the guayule plant and is formed by means of the usual retention of the 4*pro-R* hydrogen from mevalonate.

Polyprenols

The occurrence of long-chain polyisoprenoid alcohols in Nature has only been recognized fairly recently. The first metabolite that was observed with this structure, solanesol (all-*trans*-nonaprenol, (XIX)), was found in tobacco leaves and identified as a C_{45} compound with a primary hydroxyl group attached to the terminal group and associated with α,β-unsaturation.[41] Later, many closely related analogues of greater chain-length with molecular formula up to C_{120} were isolated from mammals,[42] plants,[43] yeast[44] and fungi.[45]

Mammalian tissues possess a family of polyprenols in which each molecule contains a saturated terminal α-isoprene residue (dolichol, (XX)).[42] The stereochemistry of their unsaturation has been examined with 4R- and 4S-4-$^3H_1$-labelled mevalonate. The double bonds are predominantly *cis*, in contrast with their counterparts in solanesol and phytol (in which the three saturated isoprenoid-derived residues are biogenetically *trans*[13]). In addition, hexahydropolyprenols (XXI) with extra saturated groups in the ω-terminal

$$H\left[CH_2.\overset{\overset{\displaystyle CH_3}{|}}{C}{:}CH.CH_2\right]_8.CH_2.\overset{\overset{\displaystyle CH_3}{|}}{C}{:}CH.CH_2OH$$

(XIX)

$$H\left[CH_2.\overset{\overset{\displaystyle CH_3}{|}}{C}{:}CH.CH_2\right]_{13-21}.CH_2.\overset{\overset{\displaystyle CH_3}{|}}{CH}.CH_2.CH_2OH$$

(XX)

$$CH_3.\overset{\overset{\displaystyle CH_3}{|}}{CH}.CH_2.CH_2.CH_2.\overset{\overset{\displaystyle CH_3}{|}}{CH}.CH_2.CH_2.\left[CH_2.\overset{\overset{\displaystyle CH_3}{|}}{C}{:}CH.CH_2\right]_{15-21}.CH_2.\overset{\overset{\displaystyle CH_3}{|}}{CH}.CH_2.CH_2OH$$

ω internal α

(XXI)

and adjacent isoprene positions[45] have been isolated from the mycelium of the fungus, *Aspergillus fumigatus*[46] while products that are similar in structure with solanesol, that is with unsaturation retained in all isoprene units, have been observed in plant tissues.[43] Characterization of these large molecules has been achieved after critical and detailed investigations, including microhydrogenation, mass spectrometry, nuclear magnetic resonance- and infrared-absorption spectroscopy and examination of the products of ozonolysis by chromatography.

Over half the prenols in pig liver and *A. fumigatus* exist as esters of long-chain fatty acids[47] and they are distributed fairly uniformly throughout the intracellular compartments. The free alcohols are concentrated within the mitochondria.[47,48]

These macromolecular alcohols possess intriguing structural properties but do they serve any significant purpose *in vivo*? Recent studies have established that certain polyprenols act in a cofactor capacity during transglycosidation reactions in which monosaccharide units are transferred from their water-soluble nucleotide derivatives, in the formation of polymeric material outside the confines of hydrophobic regions in the cell-membrane. A related member of this series, the C_{55} bactoprenol, has been isolated from *Lactobacillus casei*[49] and resembles the mammalian products in that it comprises of one saturated isoprene unit condensed with 10 unsaturated units. Great interest has now been aroused in undecaprenol (11 unsaturated residues) since its phosphate ester is intimately involved in diverse processes in bacteria: the biosynthesis of the mucopeptide component of cell-walls,[50] the trisaccharide moiety of the antigenic lipopolysaccharide in *Salmonella*[51] and polymannan in *Micrococcus lysodeikticus*.[52] It appears to exert its role by its ability to transfer nucleotide-linked precursors from the cytoplasm, across the cell-membrane, to the insoluble polymers that comprise the cell-wall.

Animal tissues also contain an enzyme that transfers a sugar residue to a polyprenyl (XX) phosphate acceptor. Examination of the glycolipid product formed after incubation of dolichol monophosphoglucose with liver microsomes, after solubilization with chloroform–methanol–water mixtures, showed that it consisted of an oligosaccharide with approximately 20 residues bound to the isoprene moiety through a phosphate bridge.[53]

Terpenoid Quinones and Tocopherols

Lynen[54] first proposed a mechanism based on electrophilic substitution of phenolic precursors by polyprenyl pyrophosphate, followed by loss of pyrophosphate and a proton, to explain the reaction that proceeds in the synthesis of ubiquinone, (plastoquinone), tocopherols and vitamin K (Scheme 8.9). Alkylation occurs at an early stage in the biosynthetic sequences leading to the formation of these widely-occurring products.

The polyprenyl side-chain of ubiquinone formed by *A. fumigatus* possesses an all-*trans* configuration (as in the case for squalene) and is derived after stereospecific loss of the 4*pro-S* hydrogen atom of mevalonate.[55] Similarly, measurements of $^{14}C/^{3}H$ ratios after exposure of illuminated maize shoots to [2-^{14}C, 4R-4-^{3}H$_1$]mevalonate indicated that the side-chains of plastoquinone (XXII), tocopherol, phylloquinone (vitamin K$_1$) and ubiquinone were biogenetically all-*trans* since they were formed with complete retention of the 4R tritium atom.[56]

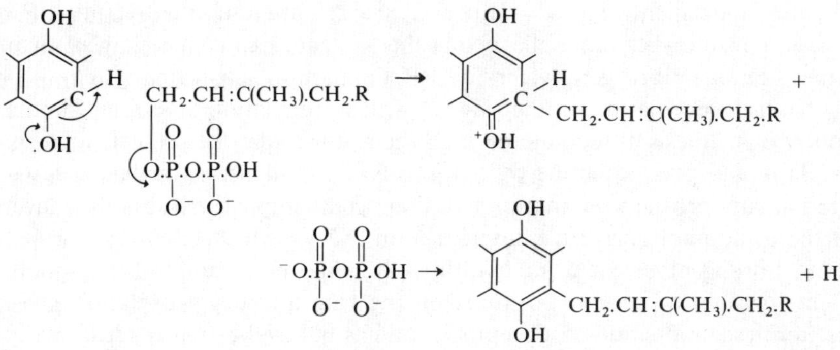

Scheme 8.9   Mechanism for the alkylation of phenols (Lynen, 1959)[54]

The first polyprenylated phenol implicated in the synthesis of ubiquinone (in *Rhodospirillum rubrum*) was characterized as 2-decaprenylphenol (XXIII).[57]

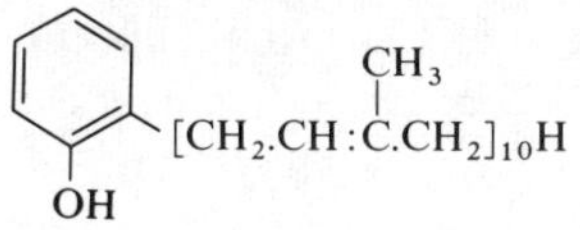

$$[CH_2.CH:C.CH_2]_{10}H$$ with CH_3 substituent, attached ortho to OH on benzene ring

(XXIII)

This lipophilic compound arises by alkylation of *p*-hydroxybenzoate with subsequent decarboxylation.[58] Later reactions yielding the fully *O*- and *C*-methylated product occur with the side-chain already attached to the aromatic nucleus.[57] Progress with cell-free studies in *E. coli* and *R. rubrum* was maintained with preparations supplemented with [^{14}C]*p*-hydroxybenzoate and an enzymic generating system for suitable polyprenyl-PP precursors.[59] The coupling activity was solubilized after sonication. The nature of the isoprenologues present in individual species reflects the preference of the alkylating enzyme for pyrophosphate esters of given chain-length. Extracts from mutants of *E. coli* that were blocked before chorismate (which cannot therefore synthesize any aromatic compound) accumulated the octaprenyl (C_{40}) side-chain.

Similar experiments were conducted with homogenates from various rat tissues and demonstrated the formation of labelled polyprenyl-*p*-hydroxybenzoate.[60] Alkylating activity resided in the mitochondria, the organelle in which ubiquinone is naturally concentrated. The aromatic ring is ultimately derived from tyrosine in animal tissues.

More recently, alkylation of added menaquinone-0 (2-methylnaphthoquinone) to menaquinone-4 (vitamin K$_2$; 4 isoprene residues) and shorter-chain derivatives has been achieved, presumably after reduction of the substrate to the corresponding phenol, by a microsomal preparation from vitamin K-deficient chicks.[61] However, exogenously supplied farnesyl-PP proved considerably more effective in this system as condensing agent than either geranyl-PP or geranylgeranyl-PP, the natural substrate.

Detailed information on the stage at which prenylation takes place in the course of synthesis of tocopherols (and their precursors, the tocotrienols, e.g. (XXIV)), and plastoquinone (XXII) is lacking. Evidence is available, however, to support the view that prenylation occurs in plant tissues at the level of the toluquinol derivative, homoarbutin (XXV), which in turn is formed from homogentisic acid and ultimately shikimic acid.[62] Prenylation, C-methylation at certain positions on the ring and deglycosylation gives rise to these lipid-soluble isoprenoid products but cyclization is additionally required for the synthesis of δ-tocotrienol (XXIV) (Scheme 8.10). Saturation of the side-chain to give a phytyl-type residue together with suitable C-methylation steps at the aromatic level then results in formation of the tocopherol series.

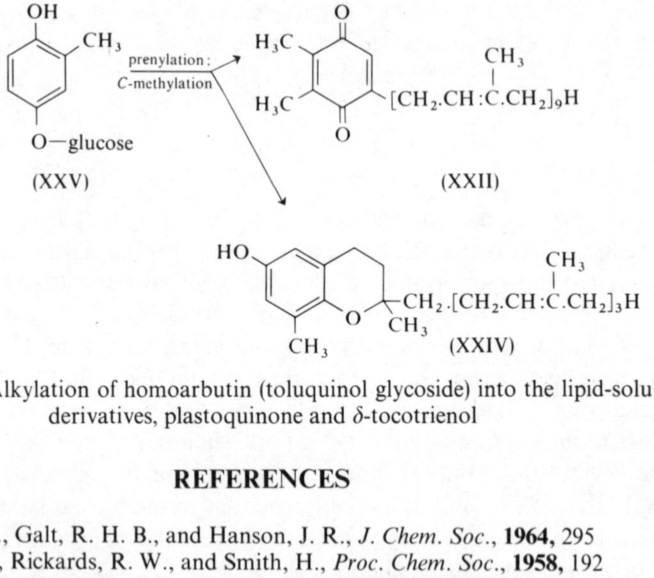

Scheme 8.10 Alkylation of homoarbutin (toluquinol glycoside) into the lipid-soluble derivatives, plastoquinone and δ-tocotrienol

REFERENCES

1. Cross, B. E., Galt, R. H. B., and Hanson, J. R., *J. Chem. Soc.*, **1964**, 295
2. Birch, A. J., Rickards, R. W., and Smith, H., *Proc. Chem. Soc.*, **1958**, 192
3. Upper, C. D., and West, C. A., *J. Biol. Chem.*, **242**, 3285 (1967)
4. Oster, M. O., and West, C. A., *Arch. Biochem. Biophys.*, **127**, 112 (1968)
5. Shechter, I., and West, C. A., *J. Biol. Chem.*, **244**, 3200 (1969)
6. Fall, R. R., and West, C. A., *J. Biol. Chem.*, **246**, 6913 (1971)
7. Graebe, J. E., Dennis, D. T., Upper, C. D., and West, C. A., *J. Biol. Chem.*, **240**, 1847 (1965)
8. Dennis, D. T., and West, C. A., *J. Biol. Chem.*, **242**, 3293 (1967)
9. Granick, S., *Annu. Rev. Plant Physiol.*, **2**, 115 (1951)
10. Lascelles, J. In *Biosynthetic Pathways in Higher Plants*, p. 163. Ed. by Pridham, J. B., and Swain, T. Academic Press Inc., London and New York, 1965
11. Wellburn, A. R., *Phytochemistry*, **9**, 2311 (1970)

12. Böger, P., *Phytochemistry*, **4**, 435 (1965)
13. Wellburn, A. R., Stone, K. J., and Hemming, F. W., *Biochem. J.*, **100**, 23c (1966)
14. Williams, R. J. H., Britton, G., Charlton, J. M., and Goodwin, T. W., *Biochem. J.*, **104**, 767 (1967)
15. Goodwin, T. W., *Adv. Enzymol.*, **21**, 295 (1959)
16. Steele, W. J., and Gurin, S., *J. Biol. Chem.*, **235**, 2778 (1960)
17. Braithwaite, G. D., and Goodwin, T. W., *Biochem. J.*, **76**, 1, 5 (1960)
18. Goodwin, T. W. In *Biosynthetic Pathways in Higher Plants*, p. 57. Ed. by Pridham, J. B., and Swain, T. Academic Press Inc., London and New York, 1965
19. Chichester, C. O., Yokoyama, H., Nakayama, T., Lukton, A., and MacKinney, G., *J. Biol. Chem.*, **234**, 598 (1959)
20. Charlton, J. M., Treharne, K. J., and Goodwin, T. W., *Biochem. J.*, **105**, 205 (1967)
21. Goodwin, T. W., and Williams, R. J. H., *Proc. Roy. Soc. Ser. B*, **163**, 515 (1966)
22. Williams, R. J. H., Britton, G. and Goodwin, T. W. *Biochem. J.*, **105**, 99 (1967)
23. Jungalwala, F. B., and Porter, J. W., *Arch. Biochem. Biophys.*, **110**, 291 (1965)
24. Jungalwala, F. B., and Porter, J. W., *Arch. Biochem. Biophys.*, **119**, 209 (1967)
25. Shah, D. V., Feldbruegge, D. H., Houser, A. R., and Porter, J. W., *Arch. Biochem. Biophys.*, **127**, 124 (1968)
26. Suzue, G., and Porter, J. W., *Biochim. Biophys. Acta*, **176**, 653 (1969)
27. Porter, J. W., and Anderson, D. G., *Annu. Rev. Plant Physiol.*, **18**, 197 (1967)
28. Porter, J. W., and Anderson, D. G., *Arch. Biochem. Biophys.*, **97**, 520 (1962)
29. Kushwaha, S. C., Subbarayan, C., Beeler, D. A., and Porter, J. W., *J. Biol. Chem.*, **244**, 3635 (1969)
30. Howes, C. D., and Batra, P. P., *Biochim. Biophys. Acta*, **222**, 174 (1970)
31. Yamamoto, H. Y., Chichester, C. O., and Nakayama, T. O. M., *Arch. Biochem. Biophys.*, **96**, 645 (1962)
32. Yamamoto, H. Y., and Chichester, C. O., *Biochim. Biophys. Acta*, **109**, 303 (1965)
33. Walton, T. J., Britton, G., and Goodwin, T. W., *Biochem. J.*, **112**, 383 (1969)
34. Goodman, D. S., Huang, H. S., and Shiratori, T., *J. Biol. Chem.*, **241**, 1929 (1966)
35. Goodman, D. S., Huang, H. S., Kanai, M., and Shiratori, T., *J. Biol. Chem.*, **242**, 3543 (1967)
36. Olson, J. A., and Hayaishi, O., *Proc. Nat. Acad. Sci. U.S.*, **54**, 1364 (1965)
36a. Wald, G., and Hubbard, R. In *The Enzymes*, Vol. 3, p. 369. Ed. by Boyer, P. D., Lardy, H., and Myrbäck, K. Academic Press Inc., London and New York, 1960
37. McMullen, A. I., and McSweeney, G. P., *Biochem. J.*, **101**, 42 (1966)
38. Cornforth, J. W., Cornforth, R. H., Popják, G., and Yengoyan, L., *J. Biol. Chem.*, **241**, 3970 (1966)
39. Archer, B. L., Barnard, D., Cockbain, E. G., Cornforth, J. W., Cornforth, R. H., and Popják, G., *Proc. Roy. Soc. Ser. B.*, **163**, 519 (1966)
40. Arreguin, B., *Handbook of Plant Physiology*, **10**, 223 (1958)
41. Erickson, R. E., Shunk, C. H., Trenner, N. R., Arison, B. H., and Folkers, K., *J. Amer. Chem. Soc.*, **81**, 4999 (1959)
42. Burgos, J., Hemming, F. W., Pennock, J. F., and Morton, R. A., *Biochem. J.*, **88**, 470 (1963)
43. Wellburn, A. R., Stevenson, J., Hemming, F. W., and Morton, R. A., *Biochem. J.*, **102**, 313 (1967)
44. Burgos, J., and Morton, R. A., *Biochem. J.*, **82**, 454 (1962)
45. Stone, K. J., Butterworth, P. H. W., and Hemming, F. W., *Biochem. J.*, **102**, 443 (1967)
46. Packter, N. M., and Glover, J., *Abst. 6th Int. Congr. Biochem. New York*, Vol. VII–113, p. 589 (1964)

47. Butterworth, P. H. W., and Hemming, F. W., *Arch. Biochem. Biophys.*, **128,** 503 (1968)
48. Stone, K. J., and Hemming, F. W., *Biochem. J.*, **109,** 877 (1968)
49. Thorne, K. J. I., and Kodicek, E., *Biochem. J.*, **99,** 123 (1966)
50. Higashi, Y., Strominger, J. L., and Sweeley, C. C., *Proc. Nat. Acad. Sci. U.S.*, **57,** 1878 (1967)
51. Wright, A., Dankert, M., Fennessey, P., and Robbins, P. W., *Proc. Nat. Acad. Sci. U.S.*, **57,** 1798 (1967)
52. Scher, M., Lennarz, W. J., and Sweeley, C. C., *Proc. Nat. Acad. Sci. U.S.*, **59,** 1313 (1968)
53. Behrens, N. H., Parodi, A. J., and Leloir, L. F., *Proc. Nat. Acad. Sci. U.S.*, **68,** 2857 (1971)
54. Lynen, F., *J. Cell. Comp. Physiol.*, **54,** Suppl. 1, 33 (1959)
55. Stone, K. J., and Hemming, F. W., *Biochem. J.*, **104,** 43 (1967)
56. Dada, O. A., Threlfall, D. R., and Whistance, G. R., *Eur. J. Biochem.*, **4,** 329 (1968)
57. Olsen, R. K., Daves, G. D., Moore, H. W., Folkers, K., Parson, W. W., and Rudney, H., *J. Amer. Chem. Soc.*, **88,** 5919 (1966)
58. Nilssen, J. L. G., Farley, T. M., and Folkers, K., *Anal. Biochem.*, **23,** 422 (1968)
59. Raman, T. S., Rudney, H., and Buzzelli, N. K., *Arch. Biochem. Biophys.*, **130,** 164 (1969)
60. Winrow, M. J., and Rudney, H., *Biochem. Biophys. Res. Commun.*, **37,** 833 (1969)
61. Dialameh, G. H., Yekundi, K. G., and Olson, R. E., *Biochim. Biophys. Acta*, **223,** 332 (1970)
62. Whistance, G. R., and Threlfall, D. R., *Biochem. J.*, **109,** 577 (1968)

Author Index

The names listed in this index refer to authors who are cited by name in the body of the text.

195

Subject Index